AF361640

Degenerative Cervical Myelopathy and the Aging Spine

Degenerative Cervical Myelopathy and the Aging Spine

Editors

Aria Nouri
Enrico Tessitore
Mark R. Kotter
Joseph S. Cheng

MDPI • Basel • Beijing • Wuhan • Barcelona • Belgrade • Manchester • Tokyo • Cluj • Tianjin

Editors
Aria Nouri
Division of Neurosurgery,
Geneva University Hospitals,
University of Geneva
Switzerland

Enrico Tessitore
Division of Neurosurgery,
Geneva University Hospitals,
University of Geneva
Switzerland

Mark R. Kotter
Honorary Consultant in
Neurosurgery, Department of
Clinical Neurosciences,
University of Cambridge
UK

Joseph S. Cheng
Department of Neurosurgery,
University of Cincinnati College
of Medicine
USA

Editorial Office
MDPI
St. Alban-Anlage 66
4052 Basel, Switzerland

This is a reprint of articles from the Special Issue published online in the open access journal *Journal of Clinical Medicine* (ISSN 2077-0383) (available at: https://www.mdpi.com/journal/jcm/special_issues/Degenerative_Cervical_Myelopathy_Aging_Spine).

For citation purposes, cite each article independently as indicated on the article page online and as indicated below:

LastName, A.A.; LastName, B.B.; LastName, C.C. Article Title. *Journal Name* **Year**, *Article Number*, Page Range.

ISBN 978-3-03943-304-9 (Hbk)
ISBN 978-3-03943-305-6 (PDF)

Contents

About the Editors . . . . **vii**

Aria Nouri, Renato Gondar, Joseph S. Cheng, Mark R.N. Kotter and Enrico Tessitore
Degenerative Cervical Myelopathy and the Aging Spine: Introduction to the Special Issue
Reprinted from: *J. Clin. Med.* **2020**, *9*, 2535, doi:10.3390/jcm9082535 . . . . **1**

Aria Nouri, Joseph S. Cheng, Benjamin Davies, Mark Kotter, Karl Schaller and Enrico Tessitore
Degenerative Cervical Myelopathy: A Brief Review of Past Perspectives, Present Developments, and Future Directions
Reprinted from: *J. Clin. Med.* **2020**, *9*, 535, doi:10.3390/jcm9020535 . . . . **9**

Daniel H. Pope, Benjamin M. Davies, Oliver D. Mowforth, A. Ramsay Bowden and Mark R. N. Kotter
Genetics of Degenerative Cervical Myelopathy: A Systematic Review and Meta-Analysis of Candidate Gene Studies
Reprinted from: *J. Clin. Med.* **2020**, *9*, 282, doi:10.3390/jcm9010282 . . . . **21**

Marco Maria Fontanella, Luca Zanin, Riccardo Bergomi, Marco Fazio, Costanza Maria Zattra, Edoardo Agosti, Giorgio Saraceno, Silvia Schembari, Lucio De Maria, Luisa Quartini, Ugo Leggio, Massimiliano Filosto, Roberto Gasparotti and Davide Locatelli
Snake-Eye Myelopathy and Surgical Prognosis: Case Series and Systematic Literature Review
Reprinted from: *J. Clin. Med.* **2020**, *9*, 2197, doi:10.3390/jcm9072197 . . . . **51**

Rocco Severino, Aria Nouri and Enrico Tessitore
Degenerative Cervical Myelopathy: How to Identify the Best Responders to Surgery?
Reprinted from: *J. Clin. Med.* **2020**, *9*, 759, doi:10.3390/jcm9030759 . . . . **63**

Stefania d'Avanzo, Marco Ciavarro, Luigi Pavone, Gabriele Pasqua, Francesco Ricciardi, Marcello Bartolo, Domenico Solari, Teresa Somma, Oreste de Divitiis, Paolo Cappabianca and Gualtiero Innocenzi
The Functional Relevance of Diffusion Tensor Imaging in Patients with Degenerative Cervical Myelopathy
Reprinted from: *J. Clin. Med.* **2020**, *9*, 1828, doi:10.3390/jcm9061828 . . . . **75**

Seok Woo Kim, Seung Bo Jang, Hyung Min Lee, Jeong Hwan Lee, Min Uk Lee, Jeong Woo Kim and Jae Sung Yee
Analysis of Cervical Spine Alignment and its Relationship with Other Spinopelvic Parameters after Laminoplasty in Patients with Degenerative Cervical Myelopathy
Reprinted from: *J. Clin. Med.* **2020**, *9*, 713, doi:10.3390/jcm9030713 . . . . **87**

Jamie R. F. Wilson, Jetan H. Badhiwala, Fan Jiang, Jefferson R. Wilson, Branko Kopjar, Alexander R. Vaccaro and Michael G. Fehlings
The Impact of Older Age on Functional Recovery and Quality of Life Outcomes after Surgical Decompression for Degenerative Cervical Myelopathy: Results from an Ambispective, Propensity-Matched Analysis from the CSM-NA and CSM-I International, Multi-Center Studies
Reprinted from: *J. Clin. Med.* **2019**, *8*, 1708, doi:10.3390/jcm8101708 . . . . **105**

Oliver Gembruch, Ramazan Jabbarli, Ali Rashidi, Mehdi Chihi, Nicolai El Hindy, Axel Wetter, Bernd-Otto Hütter, Ulrich Sure, Philipp Dammann and Neriman Özkan
Degenerative Cervical Myelopathy in Higher-Aged Patients: How Do They Benefit from Surgery?
Reprinted from: *J. Clin. Med.* **2020**, *9*, 62, doi:10.3390/jcm9010062 . . . . **117**

Insa Janssen, Aria Nouri, Enrico Tessitore and Bernhard Meyer
Cervical Myelopathy in Patients Suffering from Rheumatoid Arthritis—A Case Series of 9 Patients and A Review of the Literature
Reprinted from: *J. Clin. Med.* **2020**, *9*, 811, doi:10.3390/jcm9030811 . . . . **129**

Sukhvinder Kalsi-Ryan, Anna C. Rienmueller, Lauren Riehm, Colin Chan, Daniel Jin, Allan R. Martin, Jetan H. Badhiwala, Muhammad A. Akbar, Eric M. Massicotte and Michael G. Fehlings
Quantitative Assessment of Gait Characteristics in Degenerative Cervical Myelopathy: A Prospective Clinical Study
Reprinted from: *J. Clin. Med.* **2020**, *9*, 752, doi:10.3390/jcm9030752 . . . . **141**

Aria Nouri, Jetan H. Badhiwala, So Kato, Hamed Reihani-Kermani, Kishan Patel, Jefferson R. Wilson, Insa Janssen, Joseph S. Cheng, Karl Schaller, Enrico Tessitore and Michael G. Fehlings
The Relationship Between Gastrointestinal Comorbidities, Clinical Presentation and Surgical Outcome in Patients with DCM: Analysis of a Global Cohort
Reprinted from: *J. Clin. Med.* **2020**, *9*, 624, doi:10.3390/jcm9030624 . . . . **153**

Shreyas Panchagnula, Xin Sun, Julio D. Montejo, Aria Nouri, Luis Kolb, Justin Virojanapa, Joaquin Q. Camara-Quintana, Samuel Sommaruga, Kishan Patel, Nikita Lakomkin, Khalid Abbed and Joseph S. Cheng
Validating the Transformation of PROMIS-GH to EQ-5D in Adult Spine Patients
Reprinted from: *J. Clin. Med.* **2019**, *8*, 1506, doi:10.3390/jcm8101506 . . . . **163**

About the Editors

Aria Nouri obtained his Master of Science (under the supervision of Michael G. Fehlings) at the University of Toronto, after completing his medical training. In his thesis, he proposed and defined the term "Degenerative Cervical Myelopathy", or DCM. Thereafter, he pursued clinical research fellowships in the department of neurosurgery at Yale University and the University of Cincinnati under the mentorship of Dr. Joseph Cheng. He is currently pursuing neurosurgical residency training at Geneva University Hospitals in Switzerland.

Enrico Tessitore is the Vice-Chairman of the Department of Neurosurgery and Director of the Cancer Center at Geneva University Hospitals. Among other projects, he has been leading a prospective clinical study on degenerative cervical myelopathy, which assesses the utility of advanced MRI techniques and electrophysiological testing in diagnosing patients with DCM and predicting surgical outcome.

Mark Kotter is an honorary consultant in neurosurgery at the University of Cambridge, focusing on complex spine surgery. He is a co-founder of Myelopathy.org, and is actively engaged in both basic science and clinical research related to spinal problems. With the support of AOSpine, he has been leading RECODE-DCM, an initiative to identify and address research inefficiencies in DCM.

Joseph Cheng is the Frank H. Mayfield Chair of Neurosurgery at the University of Cincinnati, focusing on complex spine surgery. Among many noteworthy positions, he has been the chair of the AANS/CNS Spine Section, and is actively engaged in neurosurgical education and health care policy. For his continued service to neurosurgery, he received the AANS Distinguished Service Award in 2020.

Journal of
Clinical Medicine

Editorial

Degenerative Cervical Myelopathy and the Aging Spine: Introduction to the Special Issue

Aria Nouri [1,2,*]**, Renato Gondar** [1]**, Joseph S. Cheng** [2]**, Mark R.N. Kotter** [3] **and Enrico Tessitore** [1]

[1] Department of Neurosurgery, Geneva University Hospitals, 1205 Geneva, Switzerland;
rjag20@gmail.com (R.G.); Enrico.Tessitore@hcuge.ch (E.T.)

[2] Department of Neurosurgery, University of Cincinnati College of Medicine, Cincinnati, OH 45267, USA;
chengj6@ucmail.uc.edu

[3] Division of Neurosurgery, University of Cambridge, Cambridge CB2 1TN, UK; mrk25@cam.ac.uk

* Correspondence: arianouri9@gmail.com

Received: 2 August 2020; Accepted: 4 August 2020; Published: 6 August 2020

Abstract: Degenerative Cervical Myelopathy (DCM) is the most common cause of spinal cord injury in the world, but despite this, there remains many areas of uncertainty regarding the management of the condition. This special issue was dedicated to presenting current research topics in DCM. Within this issue, 12 publications are presented, including an introductory narrative overview of DCM and 11 articles comprising 9 research papers and 2 systematic reviews focusing on different aspects, ranging from genetic factors to clinical assessments, imaging, sagittal balance, surgical treatment, and outcome prediction. These articles represented contributions from a diverse group of researchers coming from multiple countries, including Switzerland, Germany, Italy, United Kingdom, United States, South Korea, and Canada.

Keywords: introduction; focus issue; spinal cord injury; compressive myelopathy; spondylosis

---

Degenerative Cervical Myelopathy (DCM) is becoming a growing public healthcare burden, attributable principally to an aging population. It represents a group of degenerative changes of the cervical spine that result in static and dynamic compression of the spinal cord, leading to subsequent chronic inflammatory and mechanical damage to neural tissue [1–3]. DCM represents the most common cause of spinal cord impairment in the developed world, leading not only to a decrease in the quality of life of those affected but also is increasingly recognized as a public healthcare and social burden [3,4]. Over the past few years, research on this topic has allowed for a better understanding of its pathological features, natural history, diagnosis, severity, associated conditions, treatment thresholds, and outcomes, collectively helping to provide a better understanding of the condition [1,5]. However, this research effort has also clearly demonstrated the ongoing knowledge gaps that exist and require further investigation [2].

The present Special Issue in the *Journal of Clinical Medicine* was specifically dedicated to presenting current research topics in DCM. Twelve articles were published, comprising 1 narrative review and 11 original articles. The narrative review focused on the past perspectives, present developments, and future directions of DCM and is intended to provide an overview of the current status of DCM [1]. The editors of the issue contributed to this introductory review and decided to limit each of the past, present, and future sections to three themes in an effort to stay focused on the most important topics. The remaining 11 articles included 9 research papers and 2 systematic reviews focusing on different aspects, ranging from genetic factors [6] to clinical assessments [7,8], imaging [9,10], sagittal balance [11], surgical treatment [12], and outcome prediction [13–16] (Table 1). These articles represented contributions from a diverse group of researchers coming from multiple countries, including Switzerland, Germany, Italy, United Kingdom, United States, South Korea, and Canada.

Several interesting findings were observed from this collective body of work. Starting with genetics, Pope et al. [6] evaluated the role of single genes in DCM onset, clinical features, and response to intervention and found genes with an effect on the radiological and clinical onset of spinal cord disease, correlating with the radiological and clinical severity of DCM. Polymorphisms of six genes were also found to have an effect on clinical response to surgery in DCM. The possible implications of this research are large, but further research will certainly be needed before this can be adapted from a clinical point of view.

Imaging-oriented research from D'Avanzo et al. [9] provided evidence that fractionated anisotropy (FA) values from MRI-DTI studies increase after decompression and potentially correlate with hand coordination and dexterity improvement, confirming previous reports that FA has an important role in prognostication. Fontanella et al. [10] reviewed the radiological finding of "snake-eye" appearance in the literature, finding some, albeit little, evidence supporting this appearance as a negative predictor. In their three illustrative cases, patients appeared to have relatively good outcomes, suggesting that further research is necessary to establish the clinical relevance of the "snake-eye" appearance. Kim et al. [11] tested the still complex and not fully understood concept of cervical sagittal balance and discussed the impact of cervical alignment on surgical decision making for laminoplasty. They concluded that the lack of kyphosis reducibility in cervical extension preoperatively should be considered as a contraindication to laminoplasty surgery.

Kalsi-Ryan et al. [7] focused on gait dysfunction in DCM, and demonstrated that DCM severity can be approximated using spatiotemporal gait parameters, providing another element of assessment that can be used to evaluate the degree and presence of functional impairment of patients. This assessment has the potential to also be used as an outcome predictor in future studies.

The majority of research focused on outcome prediction and prognostic factors. Severino et al. [13] supported the findings from D'Avanzo et al. [9], showing that FA can be used to predict surgical outcome and that increasing FA values preoperatively and postoperatively related with neurological function. Nouri at al. [14] investigated the relationship between gastrointestinal comorbidities (GICs) and DCM, as patients with GICs may suffer from anemia, inflammatory changes, and vitamin deficiencies which could be impact neurological healing. It was interesting to note that patients with and without GICs were not considerably different from a neurological function perspective, however, patients with GICs presented with a unique set of diverging characteristics including that they were more commonly female, and nearly a third of patients suffered from psychiatric comorbidities. Gembruch et al. [15] and Wilson et al. [16] focused on surgical outcomes for older patients and collectively showed that older patient clearly benefit from surgery, but may benefit less due to worse baseline neurologic function. Janssen et al. [12] reported a group of DCM patients in the context of rheumatoid arthritis, finding from their limited series that patients experience meaningful improvements in neurological function. They also showed that these patients are principally approached from a posterior approach initially.

Finally, in a topic relevant not only to the cervical spine, Panchagnula et al. [8] assessed the capacity of PROMIS to be used as a surrogate for EQ-5D, demonstrating the possibility of high accuracy mathematical transformations from PROMIS to EQ-5D in large cohorts, but limitations of accuracy of such transformation on an individual basis. This validation permits the use of PROMIS (a free questionnaire available from the NIH for the evaluation of quality of life) data to be used for quality-adjusted life year calculations.

It was the intention of this Special Issue to address a wide range of topics and we believe that the articles contained in the issue have largely achieved this objective. The editorial board also pursued this project with the hope of contributing new research to help tackle this increasingly prevalent and disabling clinical disorder. We would like to thank the various authors and peer-reviewers for helping to amass this excellent body of work.

Table 1. Summary of published papers in this Special Issue.

Authors	Purpose	Study Design	Main Results	Conclusions
D'Avanzo et al. [9]	To evaluate the use of quantitative DTI in clinical practice as a possible measure to assess clinical outcome using the mJOA and hand dexterity.	Prospective observational	FA values increase after surgery, in particular, below the most compressed level ($p = 0.044$). Postoperative FA values tend to correlate with hand dexterity ($r = 0.4272$, $R^2 = 0.0735$, $p = 0.19$ for the right hand; $r = 0.2087$, $R^2 = 0.2265$, $p = 0.53$ for the left hand), but this relationship did not show statistically significance.	FA parameters on DTI, particularly below the site of compression, may be used as a marker of myelopathy. FA increases after decompression.
Pope et al. [6]	To evaluate the role of single genes in DCM, its onset, clinical phenotype, and response to surgical intervention.	Systematic review and meta-analysis	22 genes were found to have an effect on the radiological onset of spinal column disease, while 12 influenced clinical onset of spinal cord disease. Polymorphisms of eight genes correlated with radiological severity of DCM, while three genes had an effect on clinical severity. Polymorphisms of six genes were found to have an effect on clinical response to surgery in spinal cord disease.	There are clear genetic effects on the development of spinal pathology, the central nervous system (CNS) response to bony pathology, the severity of both bony and cord pathology, and the subsequent response to surgical intervention.
Nouri et al. [1]	Provide an overview of the history of DCM (notably the transition from cervical spondylotic myelopathy to DCM), discuss current developments and interesting future directions.	Narrative review	DCM causes neurological dysfunction and is a significant cause of disability in the elderly. DCM is triggered by a variety of degenerative changes in the neck, leading to alterations in alignment, mobility, and stability, and consequently, spinal cord compression. It is a growing health problem with recently published guidelines. Many studies are currently undergoing to better direct clinical management and improve treatment outcomes.	Significant progress has been made in the field, particularly in recent years, and there are exciting possibilities for further advancements of patient care.
Panchagnula et al. [8]	To compare six health score models in a cohort of adult spine patients and to assess their ability to map PROMIS-GHS to EQ-5D in the spinal population.	Validation, prospective questionnaire	Subgroup analysis showed good predictions of the mean EQ-5D by gender, age groups, education levels, etc.	The transformation from PROMIS-GHS to EQ-5D had a high accuracy of mean estimate on a group level, but not at the individual level.
Wilson et al. [16]	To evaluate the effect of older age on the functional and QOL outcomes after surgical treatment for DCM.	Ambispective, propensity-matched analysis. International, multi-center cohort.	Significant functional improvement from the baseline was greater in the younger cohort (1-mJOA 3.8 (3.2–4.4) vs. 2.6 (2.0–3.3) $p = 0.007$; 2-SF-36 physical component summary (PCS) and mental component summary (MCS) $p \leq 0.001$, $p = 0.007$). Adverse events were not statistically significantly higher in the elderly cohort (22.4% vs. 15%; $p = 0.161$).	Elderly patients showed an improvement in functional and QOL outcomes after surgery for DCM, but the magnitude of improvement was less when compared to the matched younger adult cohort. An age over 70 was not associated with an increased risk of adverse events.

Table 1. *Cont.*

Authors	Purpose	Study Design	Main Results	Conclusions
Kim et al. [11]	To examine whether cervical alignment influences surgical outcomes.	Retrospective	Patients with a cervical lordosis had an increase in upper cervical motion (C0-2 Range of Motion (ROM), C0-2ROM/C0-7ROM) after surgery, while the non-lordosis group exhibited a decrease in C2-7ROM and C0-7ROM. Lordosis was reduced in 12 patients (22%) after surgery. All six patients belonging to the non-reducible non-lordosis group ($N = 6$) before surgery remained in the same group after the surgery.	Cervical alignment and reducibility should be identified before surgery but do not correlate with spinopelvic parameters. Lack of kyphosis reducibility in cervical extension preoperatively is a relative contraindication to laminoplasty.
Gembruch et al. [15]	To determine the surgical benefit for older (>70 years) DCM patients.	Retrospective	Preoperative and postoperative mJOA were significantly lower in patients >70 years ($p < 0.0001$). Mean mJOA improvement did not differ significantly ($p = 0.81$) six months after surgery (G1: 1.99 ± 1.04, G2: 2.01 ± 1.04, G: 2.00 ± 0.91). The delay (weeks) between symptom onset and surgery ($p = 0.003$) and the duration of the hospital stay were longer for patients >70 years old ($p < 0.0001$).	Preoperative and postoperative mJOA are affected by the patients' age, but improvement is similar. Patients should be considered for DCM surgery, regardless of their age.
Nouri et al. [14]	To investigate the difference between patients with or without gastrointestinal comorbidities (GICs) who are surgically treated for DCM.	Ambispective. International, multi-center cohort.	GICs were present in 121 patients (16%). These patients were slightly less neurologically impaired based on the Nurick grade (3.05 ± 1.10 vs. 3.28 ± 1.16, $p = 0.044$) and had a worse physical health score (32.80 ± 8.79 vs. 34.65 ± 9.38 $p = 0.049$), worse neck disability (46.31 ± 20.04 vs. 38.23 ± 20.44, $p < 0.001$), a lower prevalence of upper motor neuron signs (hyperreflexia, 70.2% vs. 78.9%, $p = 0.037$; Babinski's sign 24.8% vs. 37.3%, $p = 0.008$), and a higher rate of psychiatric comorbidities (31.4% vs. 10.4%, $p < 0.0001$). On MRI, GIC patients less commonly exhibited signal intensity changes (T2 hyperintensity, 49.2% vs. 75.6%, $p < 0.001$; T1 hypointensity, 9.7% vs. 21.1%, $p = 0.036$), and had a lower number of T2 hyperintensity levels (0.82 ± 0.98 vs. 1.3 ± 1.11, $p = 0.001$). There was no difference in surgical outcome between the groups.	DCM patients with GICs are more likely to be female and have significantly more general health impairment and neck disability, and more commonly exhibit psychiatric comorbidities. However, these patients have less clinical and MRI features typical of more severe neurological impairment.

Table 1. *Cont.*

Authors	Purpose	Study Design	Main Results	Conclusions
Kalsi-Ryan et al. [7]	To test if spatiotemporal gait parameters, including the enhanced gait variability index (eGVI), could be used to sensitively discriminate between different severities of DCM.	Prospective observational, cross-sectional	A significant correlation was found between the mJOA score and eGVI. Significant differences in the eGVI (X^2(2, $N = 153$) = 55.04, $p < 0.0001$, $\varepsilon2 = 0.36$) were found between all groups of DCM severity, with a significant increase in the eGVI as DCM progressed from mild to moderate.	The eGVI was the most discriminative gait parameter and correlated with the severity of DCM. Quantitative gait assessments are an objective tool to diagnose, classify, and evaluate the impact of therapeutic interventions in DCM.
Severino et al. [13]	To evaluate the capacity of conventional and advanced MRI techniques (using DTI), and neurophysiological parameters to identify the best candidates for decompressive surgery.	Prospective observational	There were no statistical differences in age, T2 hyperintensity, and midsagittal diameter between best and normal responders. There was a significant inverse correlation between the MEPs central conduction time and mJOA in the preoperative period ($p = 0.0004$), and a positive correlation between fractional anisotropy (FA) and mJOA during all the phases of the study, and statistically significant at 1-year (r = 0.66, $p = 0.0005$). FA was significantly higher amongst "best responders" compared to "normal responders" preoperatively and at 1-year ($p = 0.02$ and $p = 0.009$).	FA and electrophysiological aspects have a role in the diagnostic a prognostic evaluation of DCM. These results support the concept of a multidisciplinary approach in the assessment and management of DCM.
Janssen et al. [12]	To describe the rare but important presentation of cervical myelopathy in patients with rheumatoid arthritis, and its management.	Retrospective study and narrative review of literature	All patients received surgical treatment via posterior fixation, and in addition, two of these cases were combined with a transnasal anterior approach. mJOA improved from 12 ± 2.4 to 14.6 ± 1.89 at a mean follow-up at 18.8 ± 23.3 months (range 3–60 months) in five patients.	Posterior approaches are preferred for craniocervical junction instability and DCM in the context of rheumatoid arthritis. Fixation in addition to cord decompression is generally required.
Fontanella et al. [10]	To discuss the role of snake-eye appearance on MRI and its relationship with prognosis.	Case series and systematic review of the literature	Three studies which discussed snake myelopathy were reported comprising a cohort of 144 patients. "Snake-eye" appearance was regarded as a negative prognostic factor in particular, in Mizuno's study, the improvement ratio determined by JOA score was 32.2% in SEA (snake-eye appearance) vs. 47.1% in non-SEA, and 50% ($p < 0.01$) in control cases, in which high signal intensity was absent.	"Snake-eye" myelopathy represents a rare form of myelopathy and the pathophysiology is still unclear. The frequency of this presentation may be greater than previously thought and appears to be a negative prognostic factor.

J. Clin. Med. **2020**, *9*, 2535

Funding: This research received no external funding.

Conflicts of Interest: The authors declare no conflict of interest.

References

1. Nouri, A.; Cheng, J.S.; Davies, B.; Kotter, M.; Schaller, K.; Tessitore, E. Degenerative Cervical Myelopathy: A Brief Review of Past Perspectives, Present Developments, and Future Directions. *J. Clin. Med.* **2020**, *9*, 535. [CrossRef] [PubMed]
2. Tetreault, L.; Goldstein, C.L.; Arnold, P.; Harrop, J.; Hilibrand, A.; Nouri, A.; Fehlings, M.G. Degenerative Cervical Myelopathy: A Spectrum of Related Disorders Affecting the Aging Spine. *Neurosurgery* **2015**, *77* (Suppl. 4), 51–67. [CrossRef] [PubMed]
3. Nouri, A.; Tetreault, L.; Singh, A.; Karadimas, S.K.; Fehlings, M.G. Degenerative Cervical Myelopathy: Epidemiology, Genetics, and Pathogenesis. *Spine (Phila. PA 1976)* **2015**, *40*, 675–693. [CrossRef] [PubMed]
4. Nakashima, H.; Tetreault, L.A.; Nagoshi, N.; Nouri, A.; Kopjar, B.; Arnold, P.M.; Bartels, R.; Defino, H.; Kale, S.; Zhou, Q.; et al. Does age affect surgical outcomes in patients with degenerative cervical myelopathy? Results from the prospective multicenter AO Spine International study on 479 patients. *J. Neurol. Neurosurg. Psychiatry* **2016**, *87*, 734–740. [CrossRef] [PubMed]
5. Jannelli, G.; Nouri, A.; Molliqaj, G.; Grasso, G.; Tessitore, E. Degenerative Cervical Myelopathy: Review of Surgical Outcome Predictors and Need for Multimodal Approach. *World Neurosurg.* **2020**. [CrossRef] [PubMed]
6. Pope, D.H.; Davies, B.M.; Mowforth, O.D.; Bowden, A.R.; Kotter, M.R.N. Genetics of Degenerative Cervical Myelopathy: A Systematic Review and Meta-Analysis of Candidate Gene Studies. *J. Clin. Med.* **2020**, *9*, 282. [CrossRef] [PubMed]
7. Kalsi-Ryan, S.; Rienmueller, A.C.; Riehm, L.; Chan, C.; Jin, D.; Martin, A.R.; Badhiwala, J.H.; Akbar, M.A.; Massicotte, E.M.; Fehlings, M.G. Quantitative Assessment of Gait Characteristics in Degenerative Cervical Myelopathy: A Prospective Clinical Study. *J. Clin. Med.* **2020**, *9*, 752. [CrossRef] [PubMed]
8. Panchagnula, S.; Sun, X.; Montejo, J.D.; Nouri, A.; Kolb, L.; Virojanapa, J.; Camara-Quintana, J.Q.; Sommaruga, S.; Patel, K.; Lakomkin, N.; et al. Validating the Transformation of PROMIS-GH to EQ-5D in Adult Spine Patients. *J. Clin. Med.* **2019**, *8*, 1506. [CrossRef] [PubMed]
9. D'Avanzo, S.; Ciavarro, M.; Pavone, L.; Pasqua, G.; Ricciardi, F.; Bartolo, M.; Solari, D.; Somma, T.; de Divitiis, O.; Cappabianca, P.; et al. The Functional Relevance of Diffusion Tensor Imaging in Patients with Degenerative Cervical Myelopathy. *J. Clin. Med.* **2020**, *9*, 1828. [CrossRef] [PubMed]
10. Fontanella, M.; Zanin, L.; Bergomi, R.; Fazio, M.; Zattra, C.; Agosti, E.; Saraceno, G.; Schembari, S.; De Maria, L.; Quartini, L.; et al. Snake-Eye Myelopathy and Surgical Prognosis: Case Series and Systematic Literature Review. *J. Clin. Med.* **2020**, *9*, 2197. [CrossRef] [PubMed]
11. Kim, S.W.; Jang, S.B.; Lee, H.M.; Lee, J.H.; Lee, M.U.; Kim, J.W.; Yee, J.S. Analysis of Cervical Spine Alignment and its Relationship with Other Spinopelvic Parameters After Laminoplasty in Patients with Degenerative Cervical Myelopathy. *J. Clin. Med.* **2020**, *9*, 713. [CrossRef] [PubMed]
12. Janssen, I.; Nouri, A.; Tessitore, E.; Meyer, B. Cervical Myelopathy in Patients Suffering from Rheumatoid Arthritis-A Case Series of 9 Patients and A Review of the Literature. *J. Clin. Med.* **2020**, *9*, 811. [CrossRef] [PubMed]
13. Severino, R.; Nouri, A.; Tessitore, E. Degenerative Cervical Myelopathy: How to Identify the Best Responders to Surgery? *J. Clin. Med.* **2020**, *9*, 759. [CrossRef] [PubMed]
14. Nouri, A.; Badhiwala, J.H.; Kato, S.; Reihani-Kermani, H.; Patel, K.; Wilson, J.R.; Janssen, I.; Cheng, J.S.; Schaller, K.; Tessitore, E.; et al. The Relationship Between Gastrointestinal Comorbidities, Clinical Presentation and Surgical Outcome in Patients with DCM: Analysis of a Global Cohort. *J. Clin. Med.* **2020**, *9*, 624. [CrossRef] [PubMed]

15. Gembruch, O.; Jabbarli, R.; Rashidi, A.; Chihi, M.; El Hindy, N.; Wetter, A.; Hütter, B.O.; Sure, U.; Dammann, P.; Özkan, N. Degenerative Cervical Myelopathy in Higher-Aged Patients: How Do They Benefit from Surgery? *J. Clin. Med.* **2019**, *9*, 62. [CrossRef] [PubMed]
16. Wilson, J.R.F.; Badhiwala, J.H.; Jiang, F.; Wilson, J.R.; Kopjar, B.; Vaccaro, A.R.; Fehlings, M.G. The Impact of Older Age on Functional Recovery and Quality of Life Outcomes after Surgical Decompression for Degenerative Cervical Myelopathy: Results from an Ambispective, Propensity-Matched Analysis from the CSM-NA and CSM-I International, Multi-Center Studies. *J. Clin. Med.* **2019**, *8*, 1708. [CrossRef]

Journal of
Clinical Medicine

Review

Degenerative Cervical Myelopathy: A Brief Review of Past Perspectives, Present Developments, and Future Directions

Aria Nouri [1,*], Joseph S. Cheng [2], Benjamin Davies [3], Mark Kotter [3], Karl Schaller [1] and Enrico Tessitore [1]

[1] Department of Neurosurgery, University of Geneva, 1205 Geneva, Switzerland; Karl.Schaller@hcuge.ch (K.S.); enrico.tessitore@hcuge.ch (E.T.)

[2] Department of Neurosurgery, University of Cincinnati College of Medicine, Cincinnati, OH 45267-0515, USA; chengj6@ucmail.uc.edu

[3] Department of Clinical Neurosciences, University of Cambridge, Cambridge CB2 0QQ, UK; bd375@cam.ac.uk (B.D.); mrk25@cam.ac.uk (M.K.)

* Correspondence: aria.nouri@hcuge.ch; Tel.: +41-768-30-99-84

Received: 2 December 2019; Accepted: 13 February 2020; Published: 16 February 2020

Abstract: Degenerative cervical myelopathy (DCM) is the most common cause of spinal cord injury in developed countries; its prevalence is increasing due to the ageing of the population. DCM causes neurological dysfunction and is a significant cause of disability in the elderly. It has important negative impacts on the quality of life of those affected, as well as on their caregivers. DCM is triggered by a variety of degenerative changes in the neck, which affect one or more anatomical structures, including intervertebral discs, vertebrae, and spinal canal ligaments. These changes can also lead to structural abnormalities, leading to alterations in alignment, mobility, and stability. The principle unifying problem in this disease, regardless of the types of changes present, is injury to the spinal cord due to compression by static and/or dynamic forces. This review is partitioned into three segments that focus on key elements of the past, the present, and the future in the field, which serve to introduce the focus issue on "Degenerative Cervical Myelopathy and the Aging Spine". Emerging from this review is that tremendous progress has been made in the field, particularly in recent years, and that there are exciting possibilities for further advancements of patient care.

Keywords: focus issue; update; cervical spondylotic myelopathy; compressive myelopathy

1. Introduction

Degenerative cervical myelopathy (DCM) is a broad term, representing the various age-related degenerative conditions of the cervical spine that result in neurological injury to the spinal cord through static and dynamic injury mechanisms. The term DCM was introduced in 2015 in an effort to standardize the terminology [1], provide a clear definition, and provide an outline of conditions that fall under this term.

The pathogenesis of cervical spine degeneration often progresses in the following manner: general degenerative changes of the spine begin at the disc. With age, the disc becomes less compliant, principally due to a reduction in water content and fibrosis of the nucleus pulposus. This process results in the loss of the ability of the discs to distribute pressure forces equally onto the vertebral endplates. Bone remodeling at the endplates creates osteophytes and changes in the structure of vertebrae. As this process is likely triggered by pressure forces over time, the degenerative evolution may reflect a function of age and use intensity. Other alterations that occur during this process include a loss of disc and vertebral height, resulting in the in-folding of the ligamentum flavum, which may also hypertrophy in response. As a consequence of these anatomical changes, cervical alignment changes, spondylolisthesis, and hypermobility may develop. This can occur at single or multiple levels. These changes may also potentially stimulate ossification of the spinal ligaments; however, the occurrence and propensity of ossification are probably influenced by genetic factors [1–3] (Figure 1). These changes are most commonly incidental, and do not manifest symptoms, however, in some, they may become sufficient to cause spinal cord injury through static compression of the cord, dynamic injury through instability, cord stretching due to tethering, or a combination of these factors [1].

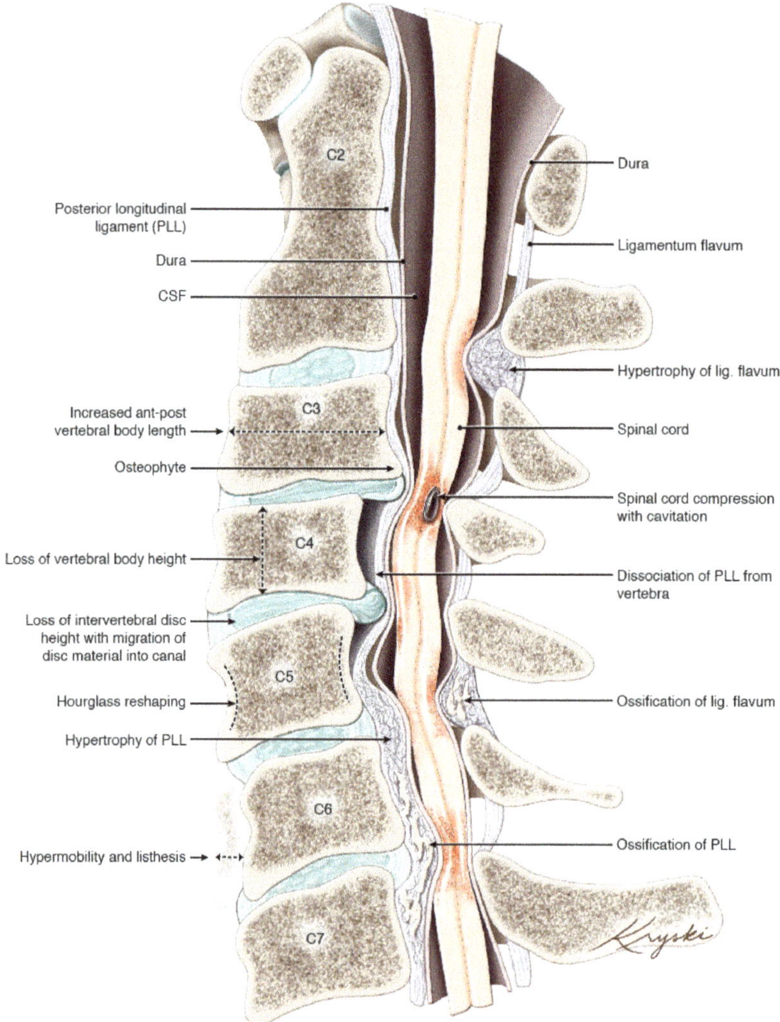

Figure 1. Artistic depiction of the various degenerative changes that can be seen in patients with DCM (Concept Aria Nouri, edits Michael G. Fehlings, artwork design Diana Kryski, copyright holder Kryski Biomedia). CSF = Cerebrospinal Fluid, PLL = Posterior Longitudinal Ligament. Originally published in Nouri et al. *Degenerative Cervical Myelopathy: Epidemiology, Genetics and Pathogenesis. Spine (Phila Pa 1976). 2015;40(12): E675-93.*

This initiates a cascade of secondary injury events within the spinal cord, including ischemia, inflammation, and apoptosis, that results in cervical myelopathy [4]. Like the degenerative pathology that causes it, the symptoms are also variable. Commonly, they can include loss of digital dexterity, weakness, imbalance and frequent falls, sensory loss, pain, and/or bladder or bowel dysfunction, in most severe cases. Together, this syndrome comprises the most common cause of spinal cord injury in the developed world.

This review will highlight some important elements of the history of DCM, classically called cervical spondylotic myelopathy (CSM), the present status, and interesting future directions. Together, this discussion serves as an introduction to the focus issue "Degenerative Cervical Myelopathy and the Aging Spine".

2. The Past

2.1. Transition from CSM to DCM

The term CSM has been used to describe vertebral degenerative disease that cause myelopathy; however, it lacked a formal and unifying definition, which has created a number of challenges. The term "spondylosis" probably came from spondylosis deformans, which itself derived from spondylitis deformans likely coined by Rokitansky in 1844 [5], mentioned by Beneke in 1897 [6], and popularized by Schmorl in 1931 [7]. The change from spondylitis to spondylosis likely arose from the distinction that spondylitis represents an infectious process, whereas spondylosis represents a degenerative process. Francois et al. (1995) discussed this history in a paper that brought together a study group from the Committee of Pathology of the European League against Rheumatism [5]. Therein, the group stated that there is no agreement on the term "spondylosis", that the group recommended avoidance of the term, and that it should be defined whenever used.

Despite this recommendation, CSM has continued to be used but variably defined. For example, ossification of the posterior longitudinal ligament (OPLL) is considered by some a subtype of CSM, and others a distinct pathology [8]. This has proved a challenge for literature synthesis [9] and has consequently hindered many important lines of investigation, including the evaluation of prevalence rates for specific phenotypes, risk factors for disease development, the natural history, and surgical decision-making. It is also possible that the inconsistent and complex terminology has contributed to a lack of disease awareness [10], which has been considered to be a contributing factor to diagnostic delay and disability [11].

Recognizing these issues, a new term, "Degenerative Cervical Myelopathy", was proposed and defined in a paper in 2015 [1]. The term encompasses both CSM and OPLL and more clearly recognizes the degenerative nature of the disease and its association with advanced age.

The transition from CSM to DCM is ongoing. However, its increasingly widespread adoption, including in the treatment guidelines by AO Spine and an ongoing research efficiency initiative [12], is indicative of its requirement and acceptance [13].

2.2. Prevention of Neurological Decline to Recovery of Neurological Function

In the past, surgical treatment was recommended to arrest further neurological decline. Historically, many even considered it a last resort. However, it has become apparent, mostly in the last decade, that many patients with DCM not only stop declining in neurological function but also may improve: in two of the largest prospective observational studies, the average improvement of neurological function based on the mJOA (modified Japanese Orthopedic Association) scale was between 2–3 [14,15]. This clarification has fundamentally altered practice, in that patients with mild myelopathy or with a stable condition may now be offered surgery, whereas before this may not have been the case. Likewise, this has also changed the counseling of patients with regards to the benefits and risks of surgery intervention.

Furthermore, there is growing evidence in the literature concerning the need for identifying patients who may benefit more from surgery [16,17]. Indeed, appropriate patient selection may lead to better surgical results, and, as a consequence, to a better appreciation of surgery among DCM patients.

Recent research, including that featured in the present special issue, indicates that improvements can occur irrespective of age. While some data indicate that old age is a predictor of a less beneficial outcome, it has consistently been shown that the elderly population is also capable of neurological recovery [18–20].

A remaining knowledge gap surrounds mild, stable myelopathy or spinal cord compression without clear or typical signs of myelopathy. Recent guidelines on this subject have indicated that patients with asymptomatic spinal cord compression should not be recommended surgery but should undergo a close clinical follow-up, and that patients with mild myelopathy should be offered surgical intervention or supervised structured rehabilitation [13]. These consensus statements are largely based on expert opinion, with limited empirical data to draw upon, owing to current practice conventions. However, the significant impact of surgery identified in these recent studies, aligned with the recognition that timely treatment is critical to recovery, may in the future see further paradigm changes.

2.3. Cervical Canal Stenosis to Cervical Cord-Canal Mismatch

The presence of a narrow canal has been widely considered to be a risk factor for the development of DCM [21]. However, the evidence is limited and largely based on data from acute traumatic spinal cord injury. This predisposition has been called congenital cervical stenosis, or developmental stenosis, or developmentally narrow canal, and measurements for this was classically done using the Torg–Pavlov ratio (TPR) or an absolute canal diameter (sometimes also called "space available for the cord" [22]) [23,24]. However, the terms used are not entirely accurate, whilst the TPR has been outdated. There may be some conditions that create a congenital stenosis, such as in achondroplasia [25], but it is unclear whether canal parameters remain the same throughout development into adulthood. Furthermore, recent studies demonstrating the measurements of normal canal and cord parameters have indicated that both the size of the canal (small canal) and the cord (large spinal cord) can predispose to the development of DCM [26]. As a consequence, spinal cord occupancy ratio (SCOR) was developed to account for both factors accurately and to enable risk determination for DCM development [27]. It has also been proposed that perhaps a cord-canal mismatch may be a better term, and that a SCOR defined as ≥70% on midsagittal imaging or ≥80% on axial imaging appears to be an effective method of identifying cord-canal mismatch [21,28,29].

Classical diagnostic criteria for diagnosing "congenital stenosis" include TPR of 0.80–0.82 [23,24] and an absolute canal diameter of 12–13 mm [30,31].

3. The Present

3.1. Current Population Trends and Epidemiology

Whilst it is well accepted that DCM is the most common cause of spinal cord injury in the developed world, the epidemiology of DCM remains poorly characterized. In North America, the incidence and prevalence was estimated at a minimum of 4.1 and 60.5 per 100,000, respectively [1]. In Taiwan, a population-based study reported a hospitalization of 4.04/100,000 person-years [32], and in the Netherlands, an incidence based on a fixed referral system of 1.6/100,000 inhabitants was reported [33]. Published studies also report that males are more commonly affected than females.

However, it is anticipated that these are significant underestimates, as they rely on operative incidence (which fails to account for the likely larger non-operative cohort) and will not account for widespread underdiagnosis. Anecdotal insights are provided by MRI series of 'healthy volunteers', which have identified an age-associated prevalence of spinal cord compression; for example, in one series of randomly selected volunteers aged 40–80, incidental cervical cord compression was detected on MRI in 59% of individuals (108/183, ranging from 31.6% in the fifth decade to 66.8% in the eighth decade) [34]. Two (1%) were found to have incidental DCM. Additionally, patients with canal stenosis or non-myelopathic compression of the spinal cord are at risk of DCM development, with those developing DCM estimated to be approximately 8% at 1-year follow-up and 23% at a median of 44-months follow-up [35]. Taken together, this data points to a significant and hidden disease burden.

The association with age aligns with emerging evidence that surgical rates for DCM are rising, particularly cervical spine fusion procedure [36–38]. It would seem that with an aging population and increased recognition of the prevalence of DCM that this trend will continue to rise. Along with this, a rise in cost and global burden should be anticipated [39].

Based on a global cohort of patients derived from the multicenter AO Spine studies on DCM, most patients present with multi-level degeneration (spondylosis) and more than 50% have accompanying ligamentum hypertrophy or in-folding that is contributing to this compression. OPLL was shown to be present in about 10% of patients, with a significantly higher prevalence in Asia [40].

3.2. Debate: Anterior vs. Posterior Surgical Approach

Decompression of the spinal cord and stabilization of the spine can be achieved safely from both anterior and posterior approaches. Furthermore, multiple options exist for either of these paths. Of course, a combined approach is also an option, but is usually reserved for the most severe cases. The relative merits of these approaches have been, and continue to be, a hot topic for debate amongst surgeons, as each approach carries differing risk profiles and healthcare costs. A recent propensity-score matched study on inpatient complications on 13,884 (n = 6942 ACDF [Anterior Cervical Decompression and Fusion]; n = 6942 PCDF [Posterior Cervical Decompression and Fusion]) patients demonstrated PCDF to be associated with greater length of stay, in-hospital costs, and general medical and surgical complications, while ACDF carried higher risk of postoperative hematoma, hoarseness, and dysphagia [41]. However, an MRI-based propensity-score-matched analysis comparing anterior vs. posterior surgery showed no significant differences in neurological outcome [42]. It is anticipated that a randomized multicenter prospective study (CSM-S trial, identifier: NCT02076113) will provide some final clarity on a subject matter that has predominated the DCM literature over the last decade [43].

Equivalence of outcome in patient populations does not mean that this also holds true at the individual level. The authors advocate a patient-tailored approach that considers individual nuances of the preexisting pathology for surgical planning; not all patients are amenable to both approaches, and management of deformity and surgical preference, as well as expertise, are important factors to consider. Indeed, this has been echoed in a global survey of spine surgeons [44], which indicated that a significant anterior cord compression and cervical kyphosis strongly influenced surgeons toward an anterior approach, whereas a high degree of posterior cord compression, congenital stenosis, and multilevel compression or OPLL were the strongest factors influencing surgeons toward a posterior surgery.

Indeed, cervical alignment has become a focus of surgical research in recent years. Baseline radiological measures including C2-7 angle, C2-7 sagittal vertical axis, T1 slope, and modified K line have been proposed to support anterior vs. posterior decision-making. For example, kyphosis exceeding 13°, based on a C2–C7 angle, has been proposed to require anterior decompression or correction of kyphosis in addition to posterior decompression [45]. Others have suggested that a high T1 slope is a risk for kyphosis development in patients with laminoplasty [46]. A more recently described measurement, the modified K-line, represents a line connecting the midpoints of the spinal cord between C2 and C7 on midsagittal MRI, which is then used to measure the minimum interval distance between the line and anterior compressive factors. Using this measurement, the authors suggest that a minimum of 4.4 mm of space between the K-line and the anterior compressive border is related with an optimal neurological recovery in non-lordotic patients after laminoplasty [47]. Taken collectively, these studies indicate alignment should be an important component of surgical decision making.

However, these studies demonstrate clinical outliers, indicating that these radiological algorithms do not fit all cases and that such measurements should represent only part of the decision-making process.

3.3. Engaging Patients through Advocacy and Knowledge Dissemination

Despite being common, there is lack of recognition with regards to DCM. The reasons for this may be multifactorial; likely contributing factors include the mentioned lack of a clear definition of CSM, the lack of advocacy or patient support, and a lack of recognition among the general medical community and beyond. However, efforts are underway to tackle this. As mentioned earlier, a new terminology has been introduced to address the lack of definition and unify the field. Furthermore, RECODE-DCM (REsearch objectives and COmmon Data Elements for DCM, https://www.recode-dcm.com) an international consensus process involving all important stakeholders (people with myelopathy, carers, surgeons, and other healthcare professionals) to (1) identify key knowledge gaps in the field, (2) define a core outcome set and data elements for DCM, and (3) reach agreement on a unifying term [12]. Moreover, we have established Myelopathy.org, the first charity for promoting DCM (Cambridge, UK). It hosts an international peer-to-peer support community and aims to improve awareness and outcomes for patients by bringing together DCM stakeholders internationally (https://myelopathy.org/).

4. The Future

4.1. Multi-Dimensional Approach to Surgical Outcome Prediction

Over the last few years, there has been a growing recognition that certain factors are able to predict surgical outcomes in patients with DCM. Prediction models using both MRI and clinical data have been developed using the global multicenter data from the AO Spine prospective studies on the effectiveness of surgery for DCM (for example, see [16,48]). However, these studies have shown that there are limits in the predictive capacity of the models. Elements that may be able to improve the current prediction models but have not yet been tested include advanced MRI and electrophysiology. Both advanced MRI measures, such as fractional anisotropy (DTI) [49], grey matter to white matter ratio (T2 *) [50], and electrophysiology [51], are, to an extent, capable of predicting surgical outcome. A novel generation of models that include such measures may improve the ability to predict surgical outcome [17]. Furthermore, machine learning techniques may also facilitate improved predictive capacities [52]. Specifically, they may provide another dimension of data analysis. However, further research is required to better understand how to utilize such techniques.

4.2. Transcranial Magnetic Stimulation (TMS) and DCM: Assessing Cortical Volume and Function

Little is known with regards to cerebral changes that occur in patients with DCM. Given that DCM patients lose physical functions, a few studies have now started to consider cortical correlates. One such recent study using TMS, a method by which a magnetic field is used to noninvasively stimulate a region of the brain, has proposed the concept of a "corticospinal reserve capacity", having identified a decrease in the cortical motor area activity and a compensatory increase in supplementary motor area activity in patients with DCM [53]. The potential implications of this concept may be significant. Does it help to explain the frequently observed discordance between clinical findings and those seen on MRI? MRI findings correlate weakly with baseline severity and outcome, and in fact patients with little compression can be disproportionately affected with regards to symptoms, whereas other patients with severe compression on imaging are relatively mildly affected with regards to their symptoms [54,55]. Indeed, differences in corticospinal reserve capacity may explain this phenomenon. Moreover, the finding, and the manner of its identification, hold therapeutic implications—is it possible, for example, to stimulate the deficient motor areas via TMS to recuperate motor function? Time will have to address this, as this area remains currently largely unexplored.

4.3. A Wave of New Biomarkers

At the present time, DCM remains a clinical diagnosis, which is confirmed via imaging, typically MRI, and sometimes with the aid of electrophysiology. Having additional biomarkers would benefit the diagnosis of DCM, and as such, this is the subject of ongoing work including advanced MRI and machine learning techniques, as outlined above. For example, quantitative MRI techniques such as grey to white matter ratio and fractional anisotropy may be able to detect progression of myelopathy severity more sensitively than classical measures such as the mJOA [56]. On the other hand, electrophysiology has been used for some time in DCM to aid diagnosis and for intraoperative monitoring purposes, but there remains significant room for improvement in this area. Some research has shown both the diagnostic and prognostic value of electrophysiology [51], however, electrophysiology has not been fully integrated into clinical practice, with the exception of its use for neuromonitoring.

Outside of imaging and electrophysiology, there are no CSF or blood biomarkers available to aid with diagnosis. However, work in this area is ongoing [57], and it is conceivable that biomarkers will be available in the future to aid in the diagnostic process. Some of this research has been done in the realm of acute spinal cord injury, which may be translatable to DCM [58], and some preliminary work has been done in DCM, such as with miR-RNA-21, miR-34a for, and miR-10a, which have been linked to neuroinflammation, neuronal apoptosis, and OPLL, respectively [59–61].

5. Conclusions

The last decade has seen significant scientific advances in DCM, driven by high quality studies in surgical management and supported by the attempt to standardize the nomenclature, including an index term. However, these studies have also demonstrated the extensive residual disability of patients after surgery and the requirement for further progress.

It is clear that an emerging goal of future research is to inform timely treatment on an individual basis, not only to arrest clinical deterioration but achieve neurological recovery. The development of multi-modal prediction models, including clinical data, imaging, and electrophysiological findings, is promising. However, it is likely to be optimized with the inclusion of new emerging diagnostic tools, such as TMS and advanced MRI techniques, and a better understanding of the biological injury.

Numerous advancements have been and are being made in the field, and ongoing research efforts promise further steps for improved patient care in DCM.

Author Contributions: Conceptualization, A.N.; Methodology; A.N.; Validation, A.N., B.D.; Investigation, A.N.; Resources, A.N., E.T.; Data curation, A.N., B.D.; Writing—original draft preparation, A.N., B.D., E.T.; writing—review and editing, A.N., J.C., B.D., M.K., K.S., E.T.; Visualization, A.N.; supervision, E.T. All authors have read and agreed to the published version of the manuscript.

Conflicts of Interest: Aria Nouri declare no conflict of interest. Enrico Tessitore declares receiving training fees from Spineart, NuVasive and DePuy Synthes. No funds were received for supporting this work.

References

1. Nouri:, A.; Tetreault, L.; Singh, A.; Karadimas, S.; Fehlings, M. Degenerative Cervical Myelopathy: Epidemiology, Genetics and Pathogenesis. *Spine* **2015**, *40*, 675–693. [CrossRef] [PubMed]
2. Galbusera, F.; van Rijsbergen, M.; Ito, K.; Huyghe, J.M.; Brayda-Bruno, M.; Wilke, H.J. Ageing and degenerative changes of the intervertebral disc and their impact on spinal flexibility. *Eur. Spine J.* **2014**, *23*, 324–332. [CrossRef] [PubMed]
3. Baptiste, D.C.; Fehlings, M.G. Pathophysiology of cervical myelopathy. *Spine J.* **2006**, *6*, 190–197. [CrossRef] [PubMed]
4. Karadimas, S.K.; Gatzounis, G.; Fehlings, M.G. Pathobiology of cervical spondylotic myelopathy. *Eur. Spine J.* **2015**, *24*, 132–138. [CrossRef] [PubMed]
5. Francois, R.J.; Eulderink, F.; Bywaters, E.G. Commented glossary for rheumatic spinal diseases, based on pathology. *Ann. Rheum. Dis.* **1995**, *54*, 615–625. [CrossRef] [PubMed]
6. Beneke, R. Zur Lehre von der Spondylitis deformans. *Vers Deutsch Katurf und Arztc* **1897**, *69*, 109–131.

7.   Schmorl, G. Beiträge zur pathologischen Anatomie der Wirbelbandscheiben und ihre Beziehungen zu den Wirbelkörpern. *Arch. Orthop. Trauma Surg.* **1931**, *29*, 389–416. [CrossRef]

8.   Davies, B.M.; McHugh, M.; Elgheriani, A.; Kolias, A.G.; Tetreault, L.; Hutchinson, P.J.; Fehlings, M.G.; Kotter, M.R. The reporting of study and population characteristics in degenerative cervical myelopathy: A systematic review. *PLoS ONE.* **2017**, *12*, e0172564. [CrossRef]

9.   Khan, D.Z.; Khan, M.S.; Kotter, M.; Davies, B. Tackling Research Inefficiency in Degenerative Cervical Myelopathy: Illustrating the current challenges for research synthesis (Preprint). *JMIR* **2019**. [CrossRef]

10.   Waqar, M.; Wilcock, J.; Garner, J.; Davies, B.; Kotter, M. Quantitative analysis of medical students' and physicians' knowledge of degenerative cervical myelopathy. *BMJ Open.* **2020**, *10*. [CrossRef]

11.   Pope, D.H.; Mowforth, O.D.; Davies, B.M.; Kotter, M.R. Diagnostic Delays Lead To Greater Disability In Degenerative Cervical Myelopathy and Represent A Health-Inequality. *Spine* **2019**. [CrossRef] [PubMed]

12.   Davies, B.M.; Khan, D.Z.; Mowforth, O.D.; McNair, A.G.; Gronlund, T.; Kolias, A.G.; Tetreault, L.; Starkey, M.L.; Sadler, I.; Sarewitz, E.; et al. RE-CODE DCM (RE search Objectives and C ommon D ata E lements for D egenerative C ervical M yelopathy): A Consensus Process to Improve Research Efficiency in DCM, Through Establishment of a Standardized Dataset for Clinical Research and the Definition of the Research Priorities. *Glob. Spine J.* **2019**, *9*, 65–76.

13.   Fehlings, M.G.; Tetreault, L.A.; Riew, K.D.; Middleton, J.W.; Aarabi, B.; Arnold, P.M.; Brodke, D.S.; Burns, A.S.; Carette, S.; Chen, R.; et al. A clinical practice guideline for the management of patients with degenerative cervical myelopathy: Recommendations for patients with mild, moderate, and severe disease and nonmyelopathic patients with evidence of cord compression. *Glob. Spine J.* **2017**, *7*, 70–83. [CrossRef] [PubMed]

14.   Fehlings, M.G.; Ibrahim, A.; Tetreault, L.; Albanese, V.; Alvarado, M.; Arnold, P.; Barbagallo, G.; Bartels, R.; Bolger, C.; Defino, H.; et al. A global perspective on the outcomes of surgical decompression in patients with cervical spondylotic myelopathy: Results from the prospective multicenter AOSpine international study on 479 patients. *Spine (Phila Pa 1976)* **2015**, *40*, 1322–1328. [CrossRef] [PubMed]

15.   Fehlings, M.G.; Wilson, J.R.; Kopjar, B.; Yoon, S.T.; Arnold, P.M.; Massicotte, E.M.; Vaccaro, A.R.; Brodke, D.S.; Shaffrey, C.I.; Smith, J.S.; et al. Efficacy and safety of surgical decompression in patients with cervical spondylotic myelopathy: Results of the AOSpine North America prospective multi-center study. *J. Bone Joint Surg. Am.* **2013**, *95*, 1651–1658. [CrossRef]

16.   Tetreault, L.A.; Kopjar, B.; Vaccaro, A.; Yoon, S.T.; Arnold, P.M.; Massicotte, E.M.; Fehlings, M.G. A clinical prediction model to determine outcomes in patients with cervical spondylotic myelopathy undergoing surgical treatment: Data from the prospective, multi-center AOSpine North America study. *J. Bone Joint Surg. Am.* **2013**, *95*, 1659–1666. [CrossRef]

17.   Tessitore, E.; Broc, N.; Mekideche, A.; Seeck, M.; Truffert, A.; Vargas, M.; Schonauer, C.; Schaller, K. A modern multidisciplinary approach to patients suffering from cervical spondylotic myelopathy. *J. Neurosurg. Sci.* **2019**, *63*, 19–29. [CrossRef]

18.   Nakashima, H.; Tetreault, L.A.; Nagoshi, N.; Nouri, A.; Kopjar, B.; Arnold, P.M.; Bartels, R.; Defino, H.; Kale, S.; Zhou, Q.; et al. Does age affect surgical outcomes in patients with degenerative cervical myelopathy? Results from the prospective multicenter AOSpine International study on 479 patients. *J. Neurol. Neurosurg. Psychiatry* **2016**, *87*, 734–740. [CrossRef]

19.   Machino, M.; Yukawa, Y.; Hida, T.; Ito, K.; Nakashima, H.; Kanbara, S.; Morita, D.; Kato, F. Can elderly patients recover adequately after laminoplasty? A comparative study of 520 patients with cervical spondylotic myelopathy. *Spine* **2012**, *37*, 667–671. [CrossRef]

20.   Lu, J.; Wu, X.; Li, Y.; Kong, X. Surgical results of anterior corpectomy in the aged patients with cervical myelopathy. *Eur. Spine J.* **2008**, *17*, 129–135. [CrossRef]

21.   Nouri, A.; Montejo, J.; Sun, X.; Virojanapa, J.; Kolb, L.E.; Abbed, K.M.; Cheng, J.S. Cervical Cord-Canal Mismatch: A New Method for Identifying Predisposition to Spinal Cord Injury. *World Neurosurg.* **2017**, *108*, 112–117. [CrossRef] [PubMed]

22.   Fujiyoshi, T.; Yamazaki, M.; Okawa, A.; Kawabe, J.; Hayashi, K.; Endo, T.; Furuya, T.; Koda, M.; Takahashi, K. Static versus dynamic factors for the development of myelopathy in patients with cervical ossification of the posterior longitudinal ligament. *J. Clin. Neurosci.* **2010**, *17*, 320–324. [CrossRef] [PubMed]

23.   Torg, J.S.; Naranja, R.J., Jr.; Pavlov, H.; Galinat, B.J.; Warren, R.; Stine, R.A. The relationship of developmental narrowing of the cervical spinal canal to reversible and irreversible injury of the cervical spinal cord in football players. *J. Bone Joint Surg. Am.* **1996**, *78*, 1308–1314. [CrossRef] [PubMed]

24. Pavlov, H.; Torg, J.S.; Robie, B.; Jahre, C. Cervical spinal stenosis: Determination with vertebral body ratio method. *Radiology* **1987**, *164*, 771–775. [CrossRef]

25. King, J.A.; Vachhrajani, S.; Drake, J.M.; Rutka, J.T. Neurosurgical implications of achondroplasia. *J. Neurosurg. Pediatr.* **2009**, *4*, 297–306. [CrossRef]

26. Nakashima, H.; Yukawa, Y.; Suda, K.; Yamagata, M.; Ueta, T.; Kato, F. Relatively Large Cervical Spinal Cord for Spinal Canal is a Risk factor for Development of Cervical Spinal Cord Compression: A Cross-Sectional Study of 1211 Subjects. *Spine (Phila Pa 1976)* **2016**, *41*, 342–348. [CrossRef]

27. Kato, F.; Yukawa, Y.; Suda, K.; Yamagata, M.; Ueta, T. Normal morphology, age-related changes and abnormal findings of the cervical spine. Part II: Magnetic resonance imaging of over 1200 asymptomatic subjects. *Eur. Spine J.* **2012**, *21*, 1499–1507. [CrossRef]

28. Nouri, A.; Tetreault, L.; Nori, S.; Martin, A.R.; Nater, A.; Fehlings, M. Congenital Cervical Spine Stenosis in a Multicenter Global Cohort of Patients with Degenerative Cervical Myelopathy: An Ambispective Report Based on a MRI Diagnostic Criterion. *Neurosurgery* **2018**, *83*, 521–528. [CrossRef]

29. Ruegg, T.B.; Wicki, A.G.; Aebli, N.; Wisianowsky, C.; Krebs, J. The diagnostic value of magnetic resonance imaging measurements for assessing cervical spinal canal stenosis. *J. Neurosurg. Spine* **2015**, *22*, 230–236. [CrossRef]

30. Lee, M.J.; Cassinelli, E.H.; Riew, K.D. Prevalence of cervical spine stenosis. Anatomic study in cadavers. *J. Bone Joint Surg. Am.* **2007**, *89*, 376–380. [CrossRef]

31. Bajwa, N.S.; Toy, J.O.; Young, E.Y.; Ahn, N.U. Establishment of parameters for congenital stenosis of the cervical spine: An anatomic descriptive analysis of 1,066 cadaveric specimens. *Eur. Spine J.* **2012**, *21*, 2467–2474. [CrossRef]

32. Wu, J.C.; Ko, C.C.; Yen, Y.S.; Huang, W.C.; Chen, Y.C.; Liu, L.; Tu, T.H.; Lo, S.S.; Cheng, H. Epidemiology of cervical spondylotic myelopathy and its risk of causing spinal cord injury: A national cohort study. *Neurosurg. Focus* **2013**, *35*, 10. [CrossRef]

33. Boogaarts, H.D.; Bartels, R.H. Prevalence of cervical spondylotic myelopathy. *Eur. Spine J.* **2015**, *24*, 139–141. [CrossRef] [PubMed]

34. Kovalova, I.; Kerkovsky, M.; Kadanka, Z.; Kadanka, Z., Jr.; Nemec, M.; Jurova, B.; Dusek, L.; Jarkovsky, J.; Bednarik, J. Author information Prevalence and imaging characteristics of nonmyelopathic and myelopathic spondylotic cervical cord compression. *Spine* **2016**, *41*, 1908–1916. [CrossRef]

35. Wilson, J.R.; Barry, S.; Fischer, D.J.; Skelly, A.C.; Arnold, P.M.; Riew, K.D.; Shaffrey, C.I.; Traynelis, V.C.; Fehlings, M.G. Frequency, timing, and predictors of neurological dysfunction in the nonmyelopathic patient with cervical spinal cord compression, canal stenosis, and/or ossification of the posterior longitudinal ligament. *Spine (Phila Pa 1976)* **2013**, *38*, 37–54. [CrossRef]

36. Lad, S.P.; Patil, C.G.; Berta, S.; Santarelli, J.G.; Ho, C.; Boakye, M. National trends in spinal fusion for cervical spondylotic myelopathy. *Surg. Neurol.* **2009**, *71*, 66–69. [CrossRef]

37. Patil, P.G.; Turner, D.A.; Pietrobon, R. National trends in surgical procedures for degenerative cervical spine disease: 1990–2000. *Neurosurgery* **2005**, *57*, 753–758. [CrossRef]

38. Marquez-Lara, A.; Nandyala, S.V.; Fineberg, S.J.; Singh, K. Current trends in demographics, practice, and in-hospital outcomes in cervical spine surgery: A national database analysis between 2002 and 2011. *Spine* **2014**, *39*, 476–481. [CrossRef]

39. Dieleman, J.L.; Squires, E.; Bui, A.L.; Campbell, M.; Chapin, A.; Hamavid, H.; Horst, C.; Li, Z.; Matyasz, T.; Reynolds, A.; et al. Factors associated with increases in US health care spending, 1996-2013. *Jama* **2017**, *318*, 1668–1678. [CrossRef]

40. Nouri, A.; Martin, A.R.; Tetreault, L.; Nater, A.; Kato, S.; Nakashima, H.; Nagoshi, N.; Reihani-Kermani, H.; Fehlings, M.G. MRI analysis of the combined prospectively collected AOSpine North America and International Data: The Prevalence and Spectrum of Pathologies in a Global Cohort of Patients with Degenerative Cervical Myelopathy. *Spine (Phila Pa 1976)* **2016**, *42*, 1058–1067. [CrossRef]

41. Badhiwala, J.H.; Ellenbogen, Y.; Khan, O.; Nouri, A.; Jiang, F.; Wilson, J.R.; Jaja, B.; Witiw, C.D.; Nassiri, F.; Fehlings, M.G.; et al. Comparison of the Inpatient Complications and Healthcare Costs of Anterior versus Posterior Cervical Decompression and Fusion in Patients with Multi-Level Degenerative Cervical Myelopathy: A Retrospective Propensity Score-Matched Analysis. *World Neurosurg.* **2019**, *1878–8750*, 32576–32578.

42. Kato, S.; Nouri, A.; Wu, D.; Nori, S.; Tetreault, L.; Fehlings, M.G. Comparison of Anterior and Posterior Surgery for Degenerative Cervical Myelopathy: An MRI-Based Propensity-Score-Matched Analysis Using Data from the Prospective Multicenter AOSpine CSM North America and International Studies. *J. Bone Joint Surg. Am.* **2017**, *99*, 1013–1021. [CrossRef] [PubMed]

43. Ghogawala, Z.; Benzel, E.C.; Heary, R.F.; Riew, K.D.; Albert, T.J.; Butler, W.E.; Barker, F.G., 2nd; Heller, J.G.; McCormick, M.P.C.; Whitmore, R.G.; et al. Cervical spondylotic myelopathy surgical trial: Randomized, controlled trial design and rationale. *Neurosurgery* **2014**, *75*, 334–346. [CrossRef] [PubMed]

44. Nouri, A.; Martin, A.R.; Nater, A.; Witiw, C.D.; Kato, S.; Tetreault, L.; Reihani-Kermani, H.; Santaguida, C.; Fehlings, M.G. The Influence of MRI Features on Surgical-Decision Making in Degenerative Cervical Myelopathy: Results From a Global Survey of AOSpine International Members. *World Neurosurg.* **2017**, *105*, 864–874. [CrossRef]

45. Suda, K.; Abumi, K.; Ito, M.; Shono, Y.; Kaneda, K.; Fujiya, M. Local kyphosis reduces surgical outcomes of expansive open-door laminoplasty for cervical spondylotic myelopathy. *Spine (Phila Pa 1976)* **2003**, *28*, 1258–1262. [CrossRef]

46. Kim, T.-H.; Lee, S.Y.; Kim, Y.C.; Park, M.S.; Kim, S.W. T1 slope as a predictor of kyphotic alignment change after laminoplasty in patients with cervical myelopathy. *Spine* **2013**, *38*, 992–997. [CrossRef]

47. Taniyama, T.; Hirai, T.; Yoshii, T.; Yamada, T.; Yasuda, H.; Saito, M.; Inose, H.; Kato, T.; Kawabata, S.; Okawa, A. Modified K-line in magnetic resonance imaging predicts clinical outcome in patients with nonlordotic alignment after laminoplasty for cervical spondylotic myelopathy. *Spine* **2014**, *39*, 1261–1268. [CrossRef]

48. Nouri, A.; Martin, A.R.; Kato, S.; Reihani-Kermani, H.; Riehm, L.E.; Fehlings, M.G. The Relationship Between MRI Signal Intensity Changes, Clinical Presentation, and Surgical Outcome in Degenerative Cervical Myelopathy: Analysis of a Global Cohort. *Spine (Phila Pa 1976)* **2017**, *42*, 1851–1858. [CrossRef]

49. Martin, A.R.; Aleksanderek, I.; Cohen-Adad, J.; Tarmohamed, Z.; Tetreault, L.; Smith, N.; Cadotte, D.W.; Crawley, A.; Ginsberg, H.; Mikulis, D.J.; et al. Translating state-of-the-art spinal cord MRI techniques to clinical use: A systematic review of clinical studies utilizing DTI, MT, MWF, MRS, and fMRI. *NeuroImage: Clinical.* **2016**, *10*, 192–238. [CrossRef]

50. Martin, A.; De Leener, B.; Cohen-Adad, J.; Cadotte, D.; Kalsi-Ryan, S.; Lange, S.; Tetreault, L.; Nouri, A.; Crawley, A.; Mikulis, D.J.; et al. A novel MRI biomarker of spinal cord white matter injury: T2*-weighted white matter to gray matter signal intensity ratio. *Am. J. Neuroradiol.* **2017**, *38*, 1266–1273. [CrossRef]

51. Holly, L.T.; Matz, P.G.; Anderson, P.A.; Groff, M.W.; Heary, R.F.; Kaiser, M.G.; Mummaneni, P.V.; Ryken, T.C.; Choudhri, T.F.; Vresilovic, E.J.; et al. Clinical prognostic indicators of surgical outcome in cervical spondylotic myelopathy. *J. Neurosurg. Spine* **2009**, *11*, 112–118. [CrossRef] [PubMed]

52. Merali, Z.G.; Witiw, C.D.; Badhiwala, J.H.; Wilson, J.R.; Fehlings, M.G. Using a machine learning approach to predict outcome after surgery for degenerative cervical myelopathy. *PLoS ONE* **2019**, *14*, e0215133. [CrossRef] [PubMed]

53. Zdunczyk, A.; Schwarzer, V.; Mikhailov, M.; Bagley, B.; Rosenstock, T.; Picht, T.; Vajkoczy, P. The corticospinal reserve capacity: Reorganization of motor area and excitability as a novel pathophysiological concept in cervical myelopathy. *Neurosurgery* **2017**, *83*, 810–818. [CrossRef] [PubMed]

54. Witiw, C.D.; Mathieu, F.; Nouri, A.; Fehlings, M.G. Clinico-Radiographic Discordance: An Evidence-Based Commentary on the Management of Degenerative Cervical Spinal Cord Compression in the Absence of Symptoms or With Only Mild Symptoms of Myelopathy. *Glob. Spine J.* **2018**, *8*, 527–534. [CrossRef]

55. Tempest-Mitchell, J.; Hilton, B.; Davies, B.M.; Nouri, A.; Hutchinson, P.J.; Scoffings, D.J.; Mannion, R.J.; Trivedi, R.; Timofeev, I.; Crawford, J.R. A comparison of radiological descriptions of spinal cord compression with quantitative measures, and their role in non-specialist clinical management. *PLoS ONE* **2019**, *14*, e0219380. [CrossRef]

56. Martin, A.R.; De Leener, B.; Cohen-Adad, J.; Kalsi-Ryan, S.; Cadotte, D.W.; Wilson, J.R.; Tetreault, L.; Nouri, A.; Crawley, A.; Mikulis, D.J. Monitoring for myelopathic progression with multiparametric quantitative MRI. *PLoS ONE* **2018**, *13*, e0195733.

57. Stewart, M.; Smith, S.; Davies, B.; Hutchinson, P.; Kotter, M. P90 A Systematic Review of Spinal Cord Serum and Cerebrospinal Fluid Biomarkers for Use in Degenerative Cervical Myelopathy. *BMJ* **2019**, *90*. [CrossRef]

58. Kwon, B.K.; Bloom, O.; Wanner, I.-B.; Curt, A.; Schwab, J.M.; Fawcett, J.; Wang, K.K. Neurochemical biomarkers in spinal cord injury. *Spinal Cord* **2019**, *57*, 819–831. [CrossRef]

59. Laliberte, A.M. An Examination of the Role of MicroRNA-21 in the Pathobiology of Degenerative Cervical Myelopathy Using Human and Animal Data. Ph.D. Thesis, University of Toronto, Toronto, ON, Canada, November 2018.
60. Xu, C.; Zhang, H.; Zhou, W.; Wu, H.; Shen, X.; Chen, Y.; Liao, M.; Liu, F.; Yuan, W. MicroRNA-10a,-210, and-563 as circulating biomarkers for ossification of the posterior longitudinal ligament. *Spine J.* **2019**, *19*, 735–743. [CrossRef]
61. Wilson, J.R.; Badhiwala, J.H.; Moghaddamjou, A.; Martin, A.R.; Fehlings, M.G. Degenerative Cervical Myelopathy; A Review of the Latest Advances and Future Directions in Management. *Neurospine* **2019**, *16*, 494. [CrossRef]

Journal of
Clinical Medicine

Review

Genetics of Degenerative Cervical Myelopathy: A Systematic Review and Meta-Analysis of Candidate Gene Studies

Daniel H. Pope [1], Benjamin M. Davies [2], Oliver D. Mowforth [1], A. Ramsay Bowden [3,4] and Mark R. N. Kotter [2,5,*]

1 School of Clinical Medicine, University of Cambridge, Cambridge CB2 0SP, UK; danielh.pope@hotmail.co.uk (D.H.P.); om283@cam.ac.uk (O.D.M.)
2 Division of Neurosurgery, Department of Clinical Neurosciences, University of Cambridge, Cambridge CB2 0QQ, UK; bd375@cam.ac.uk
3 Department of Clinical Genetics, Cambridge University Hospitals NHS Foundation Trust, Cambridge CB2 0QQ, UK; arb63@cam.ac.uk
4 The Wellcome Trust/Cancer Research UK Gurdon Institute and Department of Biochemistry, University of Cambridge, Cambridge CB2 1QN, UK
5 Anne McLaren Laboratory for Regenerative Medicine, Wellcome-MRC Cambridge Stem Cell Institute, University of Cambridge, Cambridge CB2 0SZ, UK
* Correspondence: mrk25@cam.ac.uk; Tel.: +44-122-376-3366

Received: 17 November 2019; Accepted: 14 January 2020; Published: 20 January 2020

Abstract: Degenerative cervical myelopathy (DCM) is estimated to be the most common cause of adult spinal cord impairment. Evidence that is suggestive of a genetic basis to DCM has been increasing over the last decade. A systematic search was conducted in MEDLINE, EMBASE, Cochrane, and HuGENet databases from their origin up to 14th December 2019 to evaluate the role of single genes in DCM in its onset, clinical phenotype, and response to surgical intervention. The initial search yielded 914 articles, with 39 articles being identified as eligible after screening. We distinguish between those contributing to spinal column deterioration and those contributing to spinal cord deterioration in assessing the evidence of genetic contributions to DCM. Evidence regarding a total of 28 candidate genes was identified. Of these, 22 were found to have an effect on the radiological onset of spinal column disease, while 12 genes had an effect on clinical onset of spinal cord disease. Polymorphisms of eight genes were found to have an effect on the radiological severity of DCM, while three genes had an effect on clinical severity. Polymorphisms of six genes were found to have an effect on clinical response to surgery in spinal cord disease. There are clear genetic effects on the development of spinal pathology, the central nervous system (CNS) response to bony pathology, the severity of both bony and cord pathology, and the subsequent response to surgical intervention. Work to disentangle the mechanisms by which the genes that are reviewed here exert their effects, as well as improved quality of evidence across diverse populations is required for further investigating the genetic contribution to DCM.

Keywords: genetics; single nucleotide polymorphism; degenerative cervical myelopathy; ossification posterior longitudinal ligament; severity; surgery

---

1. Introduction

Degenerative cervical myelopathy (DCM) is estimated to be the most common cause of spinal cord impairment in the adult population and its incidence is expected to rise as the population continues to age [1]. The term DCM is relatively new, and it was proposed to unify degenerative pathologies with a common injury mechanism (subacute, progressive spinal cord injury) and treatment (decompressive surgery) [1]. This includes both cervical spondylosis (such as degenerative disc disease or osteophyte formation) and the ossification of the posterior longitudinal ligament (OPLL) or ligamentum flavum (OLF) [1–4]. These aetiologies were often previously separately considered, as cervical spondylotic myelopathy (CSM) and OPLL.

The trajectory of DCM between patients is heterogenous and currently unpredictable and unexplained [3]. For example, mechanical compression is an imaging hallmark of the disease. However, the location and amount of compression does not correlate with the disease symptoms [5–7]. In fact, the clinical phenotype can range from asymptomatic to severe disability, nearly independent from the amount of compression. Furthermore, patients' response to surgical decompression, the mainstay of treatment, is variable: it achieves excellent improvements in some patients, whereas in others these do not occur [8]. Such variation between patients has led to increasing interest in the genetic basis of this condition. One study reported a relative risk of 5.21 for the development of DCM in first-degree relatives of patients [9].

So far, the effects of genes involved in inflammation, bone, and lipid metabolism have been linked to both the pathogenesis of DCM and the response to surgical intervention [10,11]. However, these studies have failed to disentangle their relationship to spinal degeneration and myelopathy. This is important, as the fact that symptom progression and severity of spinal cord compression correlate poorly suggests that the genetic polymorphisms that contribute to spinal column degeneration may be distinct from those that influence the development of myelopathy in response to the resulting spinal cord compression.

Moreover, reviews have focused on CSM or OPLL, as opposed to DCM. Genes that influence how the spinal cord copes with mechanical stress may be identifiable in studies that investigate the severity of myelopathy and, in particular, the response to surgery.

Therefore, the objectives of this review are to provide a synthesis of the published literature on a genetic contribution to the susceptibility to develop degenerative spinal column changes that lead to DCM, the heterogeneity in severity of the clinical manifestation of DCM, and the heterogeneity in response to surgery, in order to evaluate the genes that are specifically linked to the onset and recovery of myelopathy.

2. Methods

A systematic review was conducted in accordance with the PRISMA guidelines; a PRISMA checklist is presented in the Supplementary Data [12]. A search was conducted in MEDLINE, EMBASE, Cochrane, and HuGENet databases for all relevant papers from database origin up to 14th December 2019. The full search strategy is presented in the Supplementary Data and it was developed in conjunction with the Medical Library at the University of Cambridge School of Clinical Medicine. Reference lists of key articles were systematically examined to identify further eligible articles.

Titles and abstracts were screened for relevance and, subsequently, full text papers were screened for eligibility, according to the following inclusion criteria:

- Primary clinical trial
- DCM is the primary condition being addressed
- Focus on genetics (specific gene identified)
- Human study
- English language
- Full text article

Animal studies, case reports, letters, editorials, reviews, technical notes, commentaries, proposals, and corrections were excluded. In addition, articles meeting the following criteria were excluded:

- Paediatric studies (patients < 18 years)
- Focus on acute trauma and acute spinal cord injury

- Focus on thoracic or lumbar spine

Two authors independently assessed the full-texts of potentially relevant articles (DHP and BMD), with any disagreements being resolved through discussion until agreement was reached.

Data that were extracted from the eligible articles included: study design, number of cases, number of controls, participant demographics, patient disease profile, gene studied, polymorphism/haplotype studied, and effects of polymorphisms and haplotypes on DCM susceptibility/severity/response to surgery (principal summary measures: odds ratios). The risk of bias was assessed through an evaluation of study design, methods of study population selection, matching of controls to cases, and the consideration of publication source. The MINORS methodological items were used to give structure to this process [13]. The GRADE guidelines were used to rate the quality of evidence for each candidate gene, and across genes for each of the three main questions (susceptibility, severity, response) [14].

Meta-analysis using the Cochrane Review Manager 5.3 software was used for polymorphisms, where more than one study had investigated the same polymorphism and the requisite data were available.

3. Results

After removing duplicates, a total of 914 articles were screened and 39 were eligible for inclusion (Figure 1). In total, 37 articles addressed the genetics of susceptibility to developing DCM, 13 articles addressed the genetics of heterogeneity in DCM severity (either radiological or clinical severity) and six addressed the genetics of response to surgery. A total of 28 genes were identified, with key information regarding each candidate gene presented in Tables 1–3.

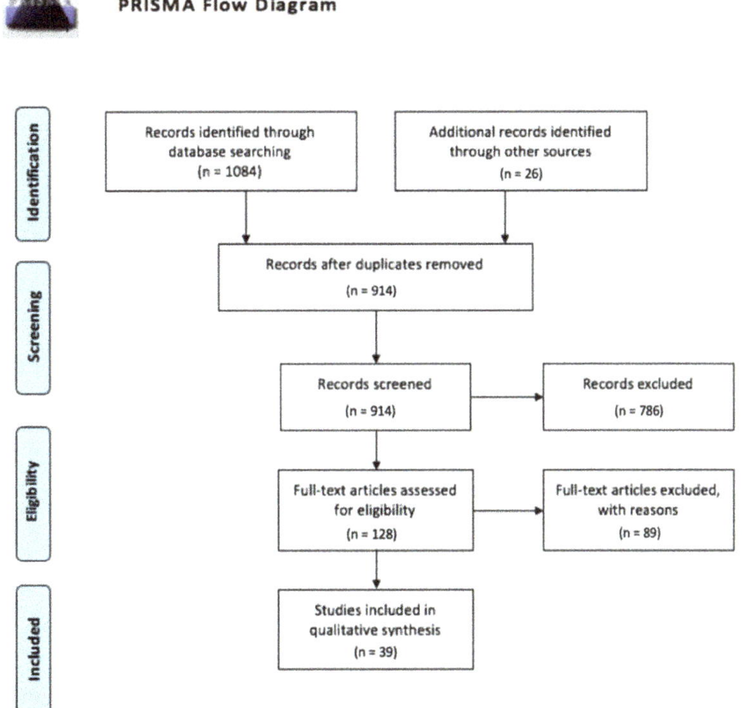

Figure 1. PRISMA flow diagram of search and screening.

3.1. What are the Genetic Effects on Susceptibility to Development of DCM?

Evidence regarding the onset of DCM/OPLL was identified for 28 genes: ACE, APOE, BID, BMP2, BMP4, BMP9, COL6A1, COL9A2, COL11A2, FGF2, FGFR1, FGFR2, HIF1A, IL1B, IL15RA, IL18RAP, leptin receptor, NPPS, OPG, OPN, RUNX2, TGFB1, TGFB3, TGFBR2, TLR5, VDBP, VDR, and VKORC1. Of these 28 genes, 22 were found to be associated with the radiological onset of spinal pathology, while 12 were associated with the clinical development of DCM (i.e., spinal cord pathology). For six genes, no significant effect of polymorphisms has been found by the studies reviewed to date: FGF2, FGFR2, IL18RAP, leptin receptor, TLR5, and VDBP. Most of the genes (19, 68%) have been investigated by only a single study. Bone morphogenetic protein genes (9, 32%) and collagen genes were the most studied gene groups (8, 29%). Table 1 presents full information for each gene.

Table 1. Susceptibility to radiological or clinical degenerative cervical myelopathy (DCM).

Candidate Gene	Papers Investigating	Study Population Location	No. of Patients	No. of Controls	Matching of Controls	Radiological or Clinical Onset of DCM	Proposed Mechanism	Odds Ratio (Susceptibility)	p-Value (Susceptibility)
ACE	Kim et al. (2014) [15]	South Korea	95 OPLL	274	Controlled for age and sex in logistic regression models	Radiological	D/D genotype	2.20	0.002
APOE	Setzer et al. (2008) [16]	Germany	60 CSM	46	Age, sex. Controls were patients with cervical spondylosis without CSM	Clinical	ε4 allele	3.50	0.008
	Diptiranhan et al. (2019) [17]	India	100 CSM	100		Clinical	ε2 allele vs. ε3 allele	4.4	0.002
							ε2 allele vs. ε4 allele	6.69	0.009
BID	Chon et al. (2014) [18]	Korea	157 OPLL	209	Controlled for age and sex in logistic regression models	Radiological	rs8190315 (Ser10 Gly) G allele	2.66	0.005
							rs2072392 (Asp60Asp) C allele	2.66	0.005
BMP2	Wang et al. (2008) [19]	China	57 OPLL	135	Age, sex	Radiological	Ser87Ser A/G allele		0.081
							Ser37Ala G allele		<0.001
	Liu et al. (2010) [20]	China	82 (48 OPLL, 12 OLF, 22 both)	118	Age, sex	Radiological	rs1005464 G allele		0.435
	Yan et al. (2013) [21]	China	420 OPLL	506	Age, sex	Radiological	109T>G G allele (Ser37Ala G allele)		<0.001
							570A>T T allele		0.005
	Kim et al. (2014) [22]	South Korea	110 OPLL	211	No. Controls were family members	Radiological	Ser87Ser A/G allele		0.411
							Ser37Ala G allele		0.670

Table 1. *Cont.*

Candidate Gene	Papers Investigating	Study Population Location	No. of Patients	No. of Controls	Matching of Controls	Radiological or Clinical Onset of DCM	Proposed Mechanism	Odds Ratio (Susceptibility)	*p*-Value (Susceptibility)
	Meng et al. (2010) [23]	China	179 OPLL	288		Radiological	−5826G>A A allele		0.495
							6007C>T T allele	1.57 (only males)	0.014
BMP4	Ren et al. (2012)a [24]	China	450 OPLL	550	Age, sex, BMI, bone mineral density, exercise level, sleeping habit, smoking status, alcohol consumption.	Radiological	rs762642 T>G G allele		0.353
							intron 2 (54422783) G>T T allele		0.868
							rs762643 C>A A allele		0.365
							rs2855530 C>G G allele		0.661
							rs2761884 C>A A allele		0.469
							intron 5 (54419501) G>A A allele		0.684
							intron 5 (54419206) C>T T allele		0.598
							intron 5 (54419150) C>T T allele	3.48	<0.001
							rs10130587 C>G G allele		0.926
							rs35107139 T>G G allele		0.953
							rs2761880 A>G G allele		0.221
							rs74486266 T>C C allele		0.861
							rs17563 C>T T allele	2.22	<0.001
							rs76335800 A>T T allele	1.99	<0.001
							3′-UTR (54416600) A>T T allele		0.190
							rs11335370 T>- deletion		0.608
							intron 6 (54416219) C>T T allele		0.344
							rs59702220 TT>- deletion		0.220
							Haplotype TGGGCTT	2.54	<0.001
	Wang et al. (2013) [25]	China	499 CSM	602	Age, sex, BMI	Clinical	−5826G>A A allele		0.214
							6007C>T T allele	0.51	<0.001

Table 1. *Cont.*

Candidate Gene	Papers Investigating	Study Population Location	No. of Patients	No. of Controls	Matching of Controls	Radiological or Clinical Onset of DCM	Proposed Mechanism	Odds Ratio (Susceptibility)	*p*-Value (Susceptibility)
BMP9	Ren et al. (2012)b [26]	China	450 OPLL	550	Age, sex, BMI, bone mineral density, exercise level, sleeping habit, smoking status, alcohol consumption.	Radiological	rs3758496		0.301
							rs12252199		0.233
							rs7923671		0.163
							rs75024165	1.82	<0.001
							rs34379100	1.95	0.003
							rs9421799	0.69	0.004
							Haplotype CTCA	2.37	<0.001
BMPR1A	Wang et al. (2018) [27]	China	356 OPLL	617	Age, sex	Radiological	−349C>T T allele		<0.001
							4A>C C allele		<0.001
							1327C>T T allele		0.311
							1395G>C		0.586

Table 1. *Cont.*

Candidate Gene	Papers Investigating	Study Population Location	No. of Patients	No. of Controls	Matching of Controls	Radiological or Clinical Onset of DCM	Proposed Mechanism	Odds Ratio (Susceptibility)	p-Value (Susceptibility)
COL6A1	Tanaka et al. (2003) [28]	Japan	342	298	Age	Radiological	rs7671 G>C allele		0.020
							rs2072699 G>A allele		0.958
							intron 2 (+758) C allele		0.019
							rs760437 C>T allele		0.435
							rs754507 A>C allele		0.062
							intron 4 (+20) C allele		0.267
							intron 4 (+37) G allele		0.010
							rs2839076 G>C allele		0.043
							intron 9 (+62) C allele		0.007
							rs2277813 C>G allele		0.057
							rs2277814 G>A allele		0.205
							rs1980982 T>C allele		0.0008
							intron 15 (+39) T allele		0.008
							rs760439 G>A allele		0.048
							rs2850173 C>A allele		0.053
							rs2075893 T>C allele		0.021
							rs2742071 T>C allele		0.219
							rs2850174 T>G allele		0.238
							rs2850175 A>C allele		0.001
							rs2839077 C>T allele		0.005
							rs2276254 A>C allele		0.00009
							rs2276255 A>G allele		0.048
							rs2276256 G>C allele		0.504
							Intron 32 (-29) C allele		0.000003
							rs2236485 G>A allele		0.0002
							rs2236486 A>G allele		0.00005
							rs2236487 A>G allele		0.00006
							rs2236488 C>T allele		0.020
							rs1053312 G>A allele		0.044
							rs1053315 G>A allele		0.040
							exon 35 (+205) T allele		0.677
							rs1053320 C>T allele		0.021
	Kong et al. (2007) [29]	China	183 (90 OPLL, 61 OLF, 32 OPLL and OLF)	155	Sex	Radiological	Promoter (−572) T allele	2.94	0.00003
							intron 32 (-29) C allele	1.89	0.004
	Liu et al. (2010) [20]	China	82 (48 OPLL, 12 OLF, 22 both)	118	Age, sex	Radiological	rs9978314 T allele		0.7618
							rs2276255 G allele		0.7354
	Kim et al. (2014) [22]	South Korea	110 OPLL	211	No. Controls were family members	Radiological	Promoter (−572) T allele		0.282
							intron 33 (+20) G allele		0.625

Table 1. *Cont.*

Candidate Gene	Papers Investigating	Study Population Location	No. of Patients	No. of Controls	Matching of Controls	Radiological or Clinical Onset of DCM	Proposed Mechanism	Odds Ratio (Susceptibility)	p-Value (Susceptibility)
COL9A2	Wang et al. (2012) [30]	China	172 CSM	176	Age, sex, BMI	Clinical	Trp2+ allele	1.78	0.048
							Trp3+ allele		0.087
COL11A2	Koga et al. (1998) [31]	Japan	124 paired siblings, 137 OPLL patients	212	No	Clinical	Promoter (−182) C allele		0.0240
							intron 6 (−4) T allele		0.0004
							exon 43 (+24) G allele		0.0210
							exon 46 (+18) T allele		0.0333
	Maeda et al. (2001) [32]	Japan	195 OPLL	187	No	Radiological	intron 6 (−4) T allele	1.99	0.0003
							exon 6 (+28) G allele	1.84	0.0012
	Horikoshi et al. (2006) [33]	Japan	711 OPLL	896	Age	Clinical	rs9277933 (IVS6-4T>A)		0.130
							rs2071025 (IVS29+37C>T)		0.270
FGF2	Jun & Kim (2012) [34]	South Korea	157 OPLL	222	Age, sex	Radiological	rs1476217 C allele		0.220
							rs308395 G allele		0.580
							rs3747676 T allele		0.100
FGFR1	Jun & Kim (2012) [34]	South Korea	157 OPLL	222	Age, sex	Radiological	rs13317 C allele	2	0.02
FGFR2	Jun & Kim (2012) [34]	South Korea	157 OPLL	222	Age, sex	Radiological	rs755793 C allele		0.110
							rs1047100 A allele		0.580
							rs3135831 T allele		0.590
HIF1A	Wang et al. (2014) [35]	China	230 CSM	284	Age, sex, BMI	Clinical	1772C>T T allele		0.760
							1790G>A A allele	1.62	<0.001
IL15RA	Kim et al. (2011) [36]	South Korea	166 OPLL	230	Age, sex	Radiological	rs2296139 A allele		1.00
							rs2228059 A allele	1.52	0.009
	Guo et al. (2014) [37]	China	235 OPLL	250	Age	Clinical	rs2296139 G allele		0.849
							rs2228059 A allele	1.63	<0.001
IL18RAP	Diptiranhan et al. (2019) [17]	India	100 CSM	100		Clinical	rs1420106		>0.05
							rs917997		>0.05
Leptin receptor	Tahara et al. (2005) [38]	Japan	156 OPLL	93	Age	Radiological	A861G		0.669

Table 1. *Cont.*

Candidate Gene	Papers Investigating	Study Population Location	No. of Patients	No. of Controls	Matching of Controls	Radiological or Clinical Onset of DCM	Proposed Mechanism	Odds Ratio (Susceptibility)	*p*-Value (Susceptibility)
NPPS	Nakamura et al. (1999) [39]	Japan	323 OPLL	332	Age	Clinical	IVS20–11delT		0.0029
	Koshizuka et al. (2002) [40]	Japan	180 OPLL	265	Age, sex	Clinical	IVS15-14T>C	3.01	0.022
	Tahara et al. (2005) [38]	Japan	156 OPLL	93	Age	Radiological	IVS20–11delT		0.512
	Horikoshi et al. (2006) [33]	Japan	711 OPLL	896	Age	Clinical	IVS15-14T>C		0.320
	He et al. (2013) [41]	China	95 OPLL	90	Age, sex	Radiological	A533C		0.430
							C973T		<0.001
							IVS15-14T>C		0.026
							IVS20–11delT		0.093
OPG	Yu et al. (2018) [42]	China	494 CSM	515		Clinical	950T>C C allele		<0.01
							1181G>C C allele		>0.05
							163A>G G allele		>0.05
OPN	Wu et al. (2014) [43]	China	187 CSM	233	Age, sex, BMI	Clinical	−66T>G G allele	1.55	0.002
							−156G/GG GG genotype		0.651
							−443C/T C allele		0.580

Table 1. *Cont.*

Candidate Gene	Papers Investigating	Study Population Location	No. of Patients	No. of Controls	Matching of Controls	Radiological or Clinical Onset of DCM	Proposed Mechanism	Odds Ratio (Susceptibility)	*p*-Value (Susceptibility)
RUNX2	Liu et al. (2010) [20]	China	82 (48 OPLL, 12 OLF, 22 both)	118	Age, sex	Radiological	rs967588C>T T allele		0.1939
							rs16873379 T>C C allele		0.169
							rs1406846 T>A A allele		0.6646
							rs3749863 A>C C allele		0.8637
							rs6908650 G>A A allele		0.6362
							rs1321075 C>A A allele		0.5255
							rs2677108 T>C C allele		0.6657
							rs16873437 G>T T allele		0.6387
							rs7771889 C>G G allele		0.7854
							rs12333172 C>T T allele		0.8128
							rs9296459 A>G G allele		0.2542
	Chang et al. (2017) [44]	China	80 OPLL	80	Age, sex, BMI, smoking history, alcohol intake	Clinical	rs967588C>T T allele	0.47	0.033
							rs16873379 T>C C allele	0.48	0.033
							rs1406846 T>A A allele	5.67	<0.001
							rs3749863 A>C C allele		0.171
							rs6908650 G>A A allele		0.959
							rs1321075 C>A A allele		0.050
							rs2677108 T>C C allele		0.295
TGFB1	Kamiya et al. (2001) [45]	Japan	46 OPLL	273	Age, BMI	Radiological	869T>C CC genotype	4.5	0.0004
	Horikoshi et al. (2006) [33]	Japan	711 OPLL	896	Age	Clinical	IVS2+114G>A A allele		0.330
	Han et al. (2013) [46]	South Korea	98 OPLL	200	Age, sex	Radiological	869T>C CC genotype		0.656
							−509C>T TT genotype		0.931
TGFB3	Horikoshi et al. (2006) [33]	Japan	711 OPLL	896	Age	Clinical	IVS1-1284G>C CC genotype	1.46	0.044
TGFBR2	Jekarl et al. (2013) [47]	South Korea	21 OPLL	42	None mentioned.	Radiological	445T>A A allele	2.81	0.007
							571G>A A allele	8.73	0.024
							1167C>T T allele		0.888

Table 1. *Cont.*

Candidate Gene	Papers Investigating	Study Population Location	No. of Patients	No. of Controls	Matching of Controls	Radiological or Clinical Onset of DCM	Proposed Mechanism	Odds Ratio (Susceptibility)	*p*-Value (Susceptibility)
TLR5	Chung et al. (2011) [48]	South Korea	166 OPLL	231	Age, sex	Radiological	rs2072493 G allele		0.457
							rs57441714 C allele		0.457
							rs5744168 T allele		0.543
VDBP	Song et al. (2018) [49]	China	318 CSM	282	Age, sex, BMI, smoking	Clinical	Thr420Lys	0.973	0.834
VDR	Kobashi et al. (2008) [50]	Japan	63 OPLL	126	Age, sex	Radiological	*FokI* FF genotype	2.33	0.0073
	Wang et al. (2010) [51]	China	154 CSM	156	Age, sex, BMI, desk work time, smoking	Clinical	*FokI* T allele		>0.05
							BsmI A allele		>0.05
							ApaI A allele	2.88	<0.001
							TaqI C allele	4.67	<0.001
	Liu et al. (2010) [20]	China	82 (48 OPLL, 12 OLF, 22 both)	118	Age, sex	Radiological	rs11168287 G allele		0.5933
							rs11574079 A allele	2.68	0.0714
							rs2189480 C allele		0.4197
							rs3847987 C allele		0.6687
							rs12721370 T allele		0.4000
	Song et al. (2018) [49]	China	318 CSM	282	Age, sex, BMI, smoking	Clinical	*FokI* FF genotype	1.461	0.001
VKORC1	Chin et al. (2013) [52]	South Korea	98 OPLL	200	Age, sex, hypertension, diabetes mellitus	Radiological	−1639G>A GA genotype	5.22 (female patients only)	0.004 (Non-significant in male/mixed)

3.1.1. Spinal Pathology

The majority of studies investigating the genetics of susceptibility to DCM used the radiological definition of cases. Therefore, these studies assess the development of bony spinal pathology (an initial stage in overall DCM development).

Kim et al. (2014) investigated the *ACE* gene, finding the deletion/deletion genotype of the intron 16 polymorphism (rs4646994) to be associated with an increased risk of developing radiological OPLL (AOR 2.20, p = 0.002) [15]. Similarly, two SNPs of the *BID* gene (rs8190315, rs2072392) were associated with the development of OPLL (OR 2.66, p = 0.005 for both) [18].

Four studies have investigated the role of variants in *BMP2*. Wang et al. (2008) found no significant effect of the Ser87Ser SNP, but found the Ser37Ala SNP was associated with an increased risk of OPLL development (p < 0.001) [19]. Interestingly, however, patients with the GG genotype of Ser87Ser had significantly greater number of ossified vertebrae, which suggested the A allele restricts ectopic ossification in OPLL. Meanwhile, the Ser37Ala SNP had no significant effect on the number of ossified vertebrae.

Yan et al. (2013) also found the Ser37Ala SNP to be associated with increased risk (p < 0.001) [21], although a more recent study that compared OPLL patients to their family members found no effect of either the Ser87Ser or Ser37Ala SNPs on risk of OPLL (p = 0.411, p = 0.670, respectively) [22]. Additionally, the 570A>T SNP in the *BMP2* gene was not found to be significantly associated with risk of OPLL [21]. Liu et al. (2010) used a patient cohort that included OPLL, OLF, and OPLL + OLF patients, but found no effect of the rs1005464 intronic SNP on the susceptibility of radiological DCM development [20].

In the *BMP4* gene, the 6007C>T SNP was found to be associated with an increased risk of developing radiological OPLL in male patients (OR 1.57, p = 0.014), although the effect is lost when males and females are considered together (p = 0.493) [23]. In the same SNP, the CT and TT genotypes were associated with a greater number of ossified vertebrae (p = 0.043) [23], as was a haplotype (TGGGCTT) containing seven SNPs (p = 0.002). Ren et al. (2012a) identified three SNPs that significantly increase the risk of OPLL: rs54419150 (OR 3.48, p < 0.001), rs17563 (OR 2.22, p < 0.001), and rs76335800 (OR 1.99, p < 0.001). Linkage disequilibrium studies also identified the haplotype block TGGGCTT containing these three SNPs to be significantly associated with the occurrence of OPLL (OR 2.54, p < 0.001) [24].

In the *BMP9* gene, two SNPs and a haplotype containing four SNPs were found to be associated with an increased risk of OPLL development: rs75024165 (OR 1.82, p < 0.001), rs34379100 (OR 1.95, p = 0.003), and haplotype CTCA (OR 2.37, p < 0.001). The haplotype was also associated with development of a greater number of ossified vertebrae (p = 0.001). A further SNP (rs9421799) was found to be protective (OR 0.69, p = 0.004), while three SNPs had no significant effect [26].

Wang et al. (2018) investigated the *BMPR1A* gene, finding two SNPs (-349C>T, 4A>C) that were associated with an increased risk of OPLL development (p < 0.001 both), and two (1327C>T, 1395G>C) with no significant effect [27]. Furthermore, patients with the C allele of the 4A>C SNP were more likely to have a greater number of ossified vertebrae on lateral cervical radiograph (p < 0.001).

The *COL6A1* gene has been the subject of four studies. Tanaka et al. (2003) investigated 32 SNPs in the *COL6A1* gene, of which 21 were significantly associated with OPLL (see Table 1) [28]. Further work by Kong et al. (2007) was consistent with these findings, with intron 32 (-29) C allele conferring a greater risk of OPLL (OR 1.89, p = 0.004) [29]. However, Liu et al. (2010) reported no significant effect of the rs2276255 SNP on the risk of OPLL or OLF development [20], in contrast to Tanaka et al.'s finding of a weak significant effect (p = 0.048). Further contradiction in the *COL6A1* gene is seen in Kong et al.'s (2007) finding that the promoter (−572) SNP T allele was associated with a 2.94 times greater risk of OPLL (p = 0.0003), while Kim et al. (2014) found no significant effect (p = 0.282) [22]. Liu et al. (2010) found no effect of one additional SNP (rs9978314) on the risk of OPLL or OLF development [20].

In the *COL11A2* gene, the intron 6 (−4) polymorphism was associated with a greater risk of OPLL development in two studies (OR 1.99, p = 0.0003; p = 0.0004) [31,32]. Similarly, the exon 6 (+28) polymorphism was associated with an odds ratio of 1.84 of developing OPLL (p = 0.0012) [32].

Jun & Kim (2012) investigated the *FGF2*, *FGFR1*, and *FGFR2* genes in 157 OPLL patients and 222 age- and sex-matched controls [34]. Three SNPs of the *FGF2* gene showed no significant effect on the likelihood of OPLL development, as did three SNPs of the *FGFR2* gene. However, the rs13317 SNP in the *FGFR1* gene was associated with an increased risk (OR 2.0, p = 0.02).

Kim et al. (2011) investigated two SNPs of the *IL15RA* (*IL15Rα*) gene [36]. The A allele of rs2228059 conferred a 1.52 times risk of radiological OPLL (p = 0.009), while the rs2296139 SNP had no significant effect.

The A861G polymorphism of the leptin receptor gene had no effect on the likelihood of OPLL development in a study of 156 OPLL patients and 93 age-matched controls [38].

In the *NPPS* gene, two studies both found no significant effect of the IVS20-11delT SNP on the likelihood of radiological OPLL (p = 0.512, p = 0.093) [38,41]. However, patients that were homozygous for the T deletion of the IVS20-11delT polymorphism had fewer ossified vertebrae and less thick ossification of their cervical vertebrae (p < 0.001 for both) [41].

The IVS15-14T>C and C973T SNPs were associated with an increased risk of radiological OPLL (p = 0.026, p < 0.001) [41]. Furthermore, patients with the T allele of the IVS15-14T>C SNP also had both a greater number of ossified vertebrae and greater thickness of ossification of their vertebrae (p < 0.001, p = 0.017, respectively). For the C973T SNP, the T allele was associated with increased thickness of ossified vertebrae (p = 0.007), but it had no effect on number of ossified vertebrae (p = 0.248). There was no effect of the A533C polymorphism on the likelihood of radiological OPLL development, or number of ossified vertebrae, or thickness of ossified vertebrae (p = 0.430, p = 0.363, p = 0.947) [41].

In a case-control study of OPLL, OLF, and OPLL+OLF patients, 11 SNPs of the *RUNX2* gene had no significant association with radiological development of OPLL/OLF [20]. However, patients with the C allele of the rs16873379 SNP had a greater number of ossified vertebrae (p = 0.001), as did patients with the A allele of the rs1406846 SNP (p = 0.020), and patients with the C allele of the rs2677108 SNP (p = 0.044).

In the *TGFB1* (*TGFβ1*) gene, the CC genotype of the 869T>C polymorphism was found to be associated with an increased risk of radiological OPLL development in one study (OR 4.5, p = 0.0004) [45], but it had no such association in a recent study that involved almost double the number of cases (p = 0.656) [46]. On meta-analysis, there was no significant effect of the 869T>C polymorphism on the susceptibility to OPLL development (OR 1.50, 95% CI 0.97–2.32, p = 0.07; Figure 2). The 509C>T was found to have no association with radiological OPLL development [46].

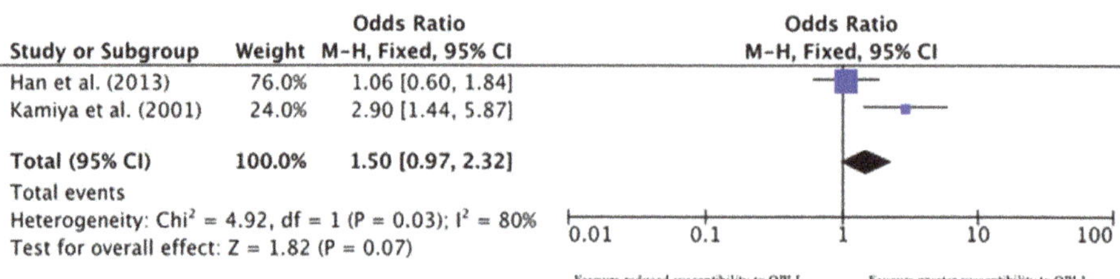

Figure 2. Forest plot for *TGFB1* 869T>C polymorphism.

Jekarl et al. (2013) investigated three SNPs of the *TGFBR2* (*TGFβR2*) gene, finding that two were associated with increased likelihood of OPLL development. The 445T>A polymorphism conferred a 2.81 times increased risk (p = 0.007), while the 571G>A polymorphism was associated with 8.73 times risk (p = 0.024) [47].

The *TLR5* gene has been investigated by one study, which found no association of three SNPs with the likelihood of OPLL development [48].

In the *VDR* gene, Kobashi et al. (2008) found the *FokI* polymorphism to be associated with 2.33 times increased risk of OPLL development ($p = 0.0073$) [50]. Similarly, Liu et al. (2010) found an association between the rs11574079 polymorphism and OPLL/OLF risk (OR 2.68, $p = 0.0714$) [20].

The *VKORC1* gene was investigated in 98 OPLL patients and 200 control subjects, with the $-1639G>$ A polymorphism having a significant effect in female patients (OR 5.22, $p = 0.004$), but not when both sexes were considered together ($p > 0.05$) [52].

In the *NPPS* gene, He et al. (2013) examined the effect of four SNPs on the progression of OPLL ossification on lateral radiograph. The AA genotype of the A533C SNP and the homozygous T deletion genotype of the IVS20-11delT SNP were both associated with better responses to surgical intervention (OR 3.11, $p = 0.029$; OR 3.35, $p = 0.007$). The other two polymorphisms were not associated with any difference in response to surgery (good response defined as <2 mm increase in ossified mass of the posterior longitudinal ligament) [41].

3.1.2. Spinal Cord Pathology

Multiple studies used clinical signs and symptoms of DCM alongside positive radiological findings. Such combination interrogates the development of cord pathology, rather than simply the development of spinal pathology.

In the *APOE* gene, the ε4 allele was found to be associated with an increased risk of myelopathy in a case-control study, where the controls had cervical spondylosis without myelopathy (OR 3.50, $p = 0.008$) [16]. However, a study in an Indian population found the ε2 allele to be associated with increased risk of myelopathy when compared to both the ε3 and ε4 alleles (OR 4.4, $p = 0.002$; OR 6.69, $p = 0.009$) [17].

In the *BMP4* gene, Wang et al. (2013) found the 6007C>T SNP to be protective for the development of clinical signs and symptoms of CSM (OR 0.51, $p < 0.001$) [25]. This is in contradiction to the evidence described above, in which this SNP was shown to be associated with an increased risk of radiological OPLL development [23,24].

The Trp2(+) allele of the *COL9A2* gene was associated with an increased risk of CSM development (OR 1.78, $p = 0.048$), a risk that was worsened by heavy smoking (OR 5.56, $p < 0.001$), while the Trp3 allele had no significant effect [30].

Koga et al. (1998) identified three polymorphisms of the *COL11A2* gene associated with DCM development: promoter (−182), exon 43 (+24) and exon 46 (+18) [31]. Horikoshi and colleagues investigated two additional SNPs of the *COL11A2* gene, but found no significant effect for either [33].

In the *HIF1A* (*HIF-1α*) gene, Wang et al. (2014) found no effect of the 1772C>T SNP, while the 1790G>A polymorphism was associated with an increased risk of CSM development (OR 1.62, $p < 0.001$) [35].

In the *IL15RA* gene, Guo et al. (2014) found a significant effect of the A allele of the rs2228059 SNP on DCM development (OR 1.63, $p < 0.001$) [37]. However, there was no effect of the rs2296139 SNP on the likelihood of developing symptomatic DCM. This is in commonality with the above findings of Kim et al. (2011) who showed rs2296139 had no effect on likelihood of developing radiological OPLL while the rs2228059 SNP did [36].

In the *IL18RAP* gene, Diptiranhan et al. (2019) found no significant effect of either the rs1420106 or rs917997 SNPs on the development of myelopathy ($p > 0.05$) [17].

Three studies have looked at the *NPPS* gene in relation to clinical onset of spinal cprd disease [33,39,40]. Nakamura et al. (1999) found the IVS20-11delT polymorphism to be associated with an increased risk of development of DCM ($p = 0.0029$) [39]. There is conflicting evidence of the effect of the IVS15-14T>C polymorphism: one study found it to be associated with a 3.01 times risk of myelopathy development ($p = 0.022$) [40], while another found no significant effect ($p = 0.320$) [33].

Yu et al. (2018) found no significant effect of the 1181G>C and 163A>G polymorphisms in the osteoprotegerin (*OPG*) gene, but found the C allele of the 950T>C SNP to be associated with a greater risk of myelopathy ($p < 0.01$) [42].

Wu et al. (2014) studied three SNPs of the osteopontin (*OPN*) gene [43]. Two showed no significant effect, but the G allele of the -66T>G SNP was associated with an odds ratio of 1.55 of clinical onset of DCM ($p = 0.002$).

In the *RUNX2* gene, Chang et al. (2017) found the SNPs rs967588 and rs16873379 to be protective for DCM development (OR 0.47, $p = 0.033$; OR 0.48, $p = 0.033$) [44]. The rs1406846 SNP was, on the other hand, strongly associated with DCM development (OR 5.67, $p < 0.001$). Four further SNPs had no significant effect.

Horikoshi et al. (2006) studied the *TGFB1* (*TGFβ1*) and *TGFB3* (*TGFβ3)* genes [33]. There was no significant effect of the IVS2+114G>A SNP of *TGFB1*, while the CC genotype IVS1-1284G>C SNP of *TGFB3* was associated with an increased risk of DCM development (OR 1.46, $p = 0.044$).

Song et al. (2018) found no significant effect of the Thr20Lys polymorphism of the *VDBP* gene (OR 0.973, $p = 0.834$) [49].

In the *VDR* gene, Wang et al. (2010) found no significant effect of *FokI* polymorphism on CSM risk [51]. The *BsmI* polymorphism also had no effect on CSM risk, but the *ApaI* and *TaqI* polymorphisms conferred a 2.88 times and 4.67 times increased CSM risk (both $p < 0.001$). In opposition to Wang et al.'s findings, Song et al. (2018) found the ff genotype of the *FokI* polymorphism to be associated with a 1.985 times greater risk of myelopathy ($p = 0.003$) [49].

3.2. What Are the Genetic Effects on Clinical Severity of DCM?

Seven studies investigated the genetic effects on the clinical severity of DCM, while 11 investigated radiological severity (four studies investigated both). Polymorphisms of 8 genes affected radiological severity, while three genes affected clinical severity. Table 2 presents the full results.

Table 2. Radiological or clinical severity of DCM.

Candidate Gene	Papers Investigating	Study Population Location	No of Patients	Method of Severity Assessment	Proposed Mechanism	Outcome
BDNF	Abode-Iyamah et al. (2016) [53]	USA	10 CSM	Short Form 36 Survey	Val66Met mutation	Met allele subjects had worse scores for physical functioning and social functioning ($p < 0.05$). Met allele subjects had worse 'physical health summary' scores ($p = 0.02$).
BMP2	Wang et al. (2008) [19]	China	57 OPLL	Number of ossified vertebrae on lateral cervical radiograph (1–7)	Ser87Ser GG genotype	Patients with GG genotype had significantly greater number of ossified vertebrae ($p < 0.001$)
					Ser37Ala GG genotype	No significant difference in number of ossified vertebrae ($p = 0.113$)
BMP4	Meng et al. (2010) [23]	China	179 OPLL	Number of ossified vertebrae on lateral cervical radiograph/CT/MRI (1–7)	−5826G>A A allele	No significant difference in number of ossified vertebrae ($p = 0.324$)
					6007C>T T allele	Patients with T allele had significantly greater number of ossified vertebrae ($p = 0.043$)
	Ren et al. (2012)a [24]	China	450 OPLL	Number of ossified vertebrae on lateral cervical radiograph/CT (1–7)	Haplotype TGGGCTT	Patients with the TGGGCTT haplotype had significantly greater number of ossified vertebrae ($p = 0.002$)
BMP9	Ren et al. (2012)b [26]	China	450 OPLL	Number of ossified vertebrae on lateral cervical radiograph/CT (1–7)	Haplotype CTCA	Patients with the CTCA haplotype had significantly greater number of ossified vertebrae ($p = 0.001$)
BMPR1A	Wang et al. (2018) [27]	China	356 OPLL	Number of ossified vertebrae on lateral cervical radiograph (1–7)	4A>C C allele	Patients with C allele had significantly greater number of ossified vertebrae ($p < 0.001$)
HIF1A	Wang et al. (2014) [35]	China	230 CSM	mJOA score	1772C>T T allele	No significant difference in mJOA score ($p > 0.05$)
					1790G>A A allele	Patients with A allele had significantly worse mJOA scores ($p < 0.001$)
NPPS	He et al. (2013) [41]	China	95 OPLL	Number of ossified vertebrae on lateral cervical radiograph (1–7)	A533C	No significant difference in number of ossified vertebrae ($p = 0.363$)
					C973T	No significant difference in number of ossified vertebrae ($p = 0.248$)
					IVS15–14T>C	Patients with T allele had significantly greater number of ossified vertebrae ($p < 0.001$)
					IVS20–11delT	Patients homozygous for the T deletion had significantly fewer ossified vertebrae ($p < 0.001$)
				Ossified thickness of cervical vertebrae on lateral radiograph	A533C	No significant difference in ossified thickness of cervical vertebrae ($p = 0.947$)
					C973T	Patients with T allele had significantly thicker ossification of cervical vertebrae ($p = 0.007$)
					IVS15–14T>C	Patients with T alelle had significantly thicker ossification of cervical vertebrae ($p = 0.017$)
					IVS20–11delT	Patients homozygous for the T deletion had significantly less thick ossification of cervical vertebrae ($p < 0.001$)
OPG	Yu et al. (2018) [42]	China	494 CSM	mJOA score and number of ossified vertebrae	950T>C	TT genotype associated with higher mJOA scores and fewer ossified cervical vertebrae ($p < 0.05$).

Table 2. *Cont.*

Candidate Gene	Papers Investigating	Study Population Location	No of Patients	Method of Severity Assessment	Proposed Mechanism	Outcome
OPN	Wu et al. (2014) [43]	China	187 CSM	mJOA score	−66T>G G allele	No significant difference in mJOA score ($p > 0.05$)
					−156G/GG GG genotype	No significant difference in mJOA score ($p > 0.05$)
					−443C/T C allele	No significant difference in mJOA score ($p > 0.05$)
RUNX2	Chang et al. (2017) [44]	China	80 OPLL	Number of ossified vertebrae on CT/MRI (1–7)	rs967588C>T T allele	No significant difference in number of ossified vertebrae ($p = 0.784$)
					rs16873379 T>C C allele	Patients with C allele had significantly greater number of ossified vertebrae ($p = 0.001$)
					rs3749863 A>C C allele	No significant difference in number of ossified vertebrae ($p = 0.129$)
					rs6908650 G>A A allele	No significant difference in number of ossified vertebrae ($p = 0.813$)
					rs1321075 C>A A allele	No significant difference in number of ossified vertebrae ($p = 0.610$)
					rs1406846 T>A A allele	Patients with A allele had significantly greater number of ossified vertebrae ($p = 0.020$)
					rs2677108 T>C C allele	Patients with C allele had significantly greater number of ossified vertebrae ($p = 0.044$)
VDBP	Song et al. (2018) [49]	China	318 CSM	mJOA score	Thr420Lys	No significant difference in mJOA score ($p = 0.546$)
				Number of ossified vertebrae	Thr420Lys	No significant difference in number of ossified vertebrae ($p = 0.117$)
VDR	Wang et al. (2010) [51]	China	154 CSM	Number of segmental lesions on MRI	*FokI* T allele	No significant difference in mean number of segmental lesions ($p > 0.05$)
					BsmI A allele	No significant difference in mean number of segmental lesions ($p > 0.05$)
					ApaI A allele	No significant difference in mean number of segmental lesions ($p > 0.05$)
					TaqI C allele	No significant difference in mean number of segmental lesions ($p > 0.05$)
				mJOA score	*FokI* T allele	No significant difference in mJOA score ($p > 0.05$)
					BsmI A allele	No significant difference in mJOA score ($p > 0.05$)
					ApaI A allele	No significant difference in mJOA score ($p > 0.05$)
					TaqI C allele	No significant difference in mJOA score ($p > 0.05$)
	Song et al. (2018) [49]	China	318 CSM	mJOA score	*FokI* ff genotype	No significant difference in mJOA score ($p = 0.358$)
				Number of ossified vertebrae	*FokI* ff genotype	No significant difference in number of ossified vertebrae ($p = 0.575$)

CSM patients with the Val66Met polymorphism of the *BDNF* gene had more severe disease, as assessed by functional survey: worse SF-36 scores for physical functioning and physical health summary than their counterparts without the polymorphism ($p < 0.05$) [53].

Wang et al. (2014) studied the effect of two polymorphisms of the *HIF1A* gene on CSM: 1772C>T and 1790G>A [35]. While the former conferred no significant difference in mJOA score, in the latter patients with the A allele had significantly worse mJOA scores than their G allele counterparts ($p < 0.001$).

Yu et al. (2018) found the TT genotype of the 950T>C polymorphism in the *OPG* gene to be associated with higher mJOA scores and fewer ossified vertebrae ($p < 0.05$); the TT genotype appears to be protective [42].

Wu et al. (2014) investigated four polymorphisms of the *OPN* gene in 187 CSM patients, finding no significant difference of all four polymorphisms on the mJOA score [43].

There was no effect of the Thr420Lys polymorphism of the *VDBP* gene on mJOA score or the number of ossified segments in 318 CSM patients [49]. Similarly, four polymorphisms of the *VDR* gene (*FokI, BsmI, ApaI, TaqI*) were found to have no significant effect on mJOA score in two studies [49,51].

3.3. What Are the Genetic Effects on Response to Surgery in DCM?

The polymorphisms of five genes were associated with clinical response to surgery in DCM: *APOE, BMP4, HIF1A, OPN,* and *RUNX2*. The *NPPS* gene was studied for radiological response to surgery. Table 3 presents the results.

In the *APOE* gene, the ε4 allele was associated with an increased risk of poor response to ACDF surgery. In a multivariate model, it was associated with an 8.6 times risk of worsening or no change in mJOA score ($p = 0.004$) [54].

The 6007C>T polymorphism of the *BMP4* gene was associated with greater likelihood of post-surgical improvement of mJOA score (OR 1.53, $p = 0.002$), but the -5826G>A polymorphism had no significant effect ($p = 0.053$) [25].

In the *HIF1A* gene, the 1790G>A polymorphism was also associated with a greater likelihood of post-surgical improvement of the mJOA score (OR 1.55, $p = 0.024$) [35].

In the *OPN* gene, the GG genotype of the −66T>G SNP was found to be associated with worse response to surgical intervention, as assessed by mJOA score (OR 3.62, $p = 0.007$) [43]. Good surgical response was defined as >50% improvement in mJOA score.

Seven polymorphisms of the *RUNX2* gene were investigated for their effect on pre- vs. post-surgical mJOA score. The patients with the CC genotype of the rs16873379 SNP improved less (52.4%) than patients with TT genotype (61.7%), an effect that is mirrored by patients with the AA genotype of the rs1406846 SNP and patients with the CC genotype of the rs2677108 SNP. Patients with the AA genotype of the rs6908650 SNP improved more (66.8%) than their counterparts with the GG genotype (57.4%). The three other polymorphisms had no significant effect on mJOA score improvement [44].

In the NPPS gene, the AA genotype of the A533C polymorphism was associated with a 3.11 times greater likelihood of radiological improvement after surgical intervention. Similarly the IVS20-11delT homozygous T deletion was associated with a 3.35 greater likelihood of improvement. For both polymorphisms, improvement was defined as an increase of <2 mm in the ossified mass of the posterior longitudinal ligament over a mean follow-up length of 3.1 years [41].

Table 3. Response to surgery in DCM.

Candidate Gene	Papers Investigating	Study Population Location	Number of Patients	Surgery Type	Mean Follow-Up	Method of Assessment of Response to Surgery	Improvement Defined As	Proposed Mechanism	Odds Ratio of No Improvement	Odds Ratio of Improvement	p-Value
APOE	Setzer et al. (2009) [54]	Germany	60 CSM	ACDF	18.8 months	mJOA score	mJOA score +1	ε4 allele	3.3 (8.6 in multivariate model)	-	**0.002** (0.004 multivariate model)
BMP4	Wang et al. (2013) [25]	China	499 CSM	Anterior cervical corpectomy and fusion	12 months	mJOA score	>50% improvement in mJOA score	−5826G>A A allele	-	-	0.053
								6007C>T T allele	-	1.53	**0.002**
HIF1A	Wang et al. (2014) [35]	China	230 CSM	Anterior cervical corpectomy and fusion	24 months	mJOA score	>50% improvement in mJOA score	1790G>A A allele	-	1.55	**0.024**
NPPS	He et al. (2013) [41]	China	95 OPLL		3.1 years	Progression of OPLL ossification on lateral radiograph	<2 mm increase in ossified mass of PLL	A533C AA genotype	-	3.11	**0.029**
								C973T	-	-	0.935
								IVS15-14T>C	-	-	0.836
								IVS20–11delT homozygous T deletion	-	3.35	**0.007**
OPN	Wu et al. (2014) [43]	China	187 CSM	Anterior cervical corpectomy and fusion	24 months	mJOA score	>50% improvement in mJOA score	−66T>G GG genotype	3.62	-	**0.007**
RUNX2	Chang et al. (2017) [44]	China	80 OPLL	Laminoplasty	12 months	mJOA score	% improvement in mJOA score	rs967588C>T T allele	-	-	>0.05
								rs16873379 T>C C allele	-	-	**<0.05**
								rs3749863 A>C C allele	-	-	>0.05
								rs6908650 G>A A allele	-	-	**<0.05**
								rs1321075 C>A A allele	-	-	>0.05
								rs1406846 T>A A allele	-	-	**<0.05**
								rs2677108 T>C C allele	-	-	**<0.05**

4. Discussion

The aim of this study was to critically appraise the current evidence on the genetic contribution to DCM, with specific focus on distinguishing spinal column disease from spinal cord disease. Studies were identified evaluating the susceptibility, severity, and responsiveness to surgery in DCM. Studies on spinal column disease focused on the radiological outcomes of OPLL. Evidence was identified for a number of genes, including many in the *TGFβ* superfamily and many known to be associated with bone development.

By further focusing on studies evaluating relationships with clinical function, versus radiological measures, a shortlist of genes that were related to spinal column disease or 'myelopathy' and not 'spondylosis' was identified: specifically, 12 genes that were associated with susceptibility, three genes with clinical severity, and five genes with response to surgical intervention. Table 4 presents a summary of the evidence for genetic effects on 'myelopathy', including GRADE rating for each gene. Across the three focuses of this review (susceptibility, severity, response to surgery), the GRADE rating of quality of evidence is baseline low, as all studies are observational. For all three, the quality of evidence is upgraded due to the large effects across genes, but downgraded due to inconsistency between studies.

Table 4. Summary of candidate genes affecting myelopathy (i.e., *clinical* onset/severity/response to surgery rather than radiological). Colour coded for evidence level (green: unconflicted evidence, amber: conflicting evidence, red: no evidence or not yet investigated). GRADE rating of quality of evidence given for each candidate gene—baseline quality low (all observational studies); gene-specific upgrade/downgrade comments in parentheses.

Candidate Gene	Papers Investigating	Susceptibility to Myelopathy	Severity of Myelopathy	Post-Operative Response	GRADE Rating
APOE	Setzer et al. (2008) [16] Setzer et al. (2009) [54]	ε4 allele: OR 3.50, $p = 0.008$		ε4 allele: OR of no improvement 3.3 (8.6 in multivariate model), $p = 0.002$ ($p = 0.004$)	Low (small sample size, inconsistency across ethnicities)
	Diptiranhan et al. (2019)	ε2 allele: OR 6.69, $p = 0.009$			
BDNF	Abode-Iyamah et al. (2016) [53]		Val66Met: Met allele subjects had worse scores for physical functioning ($p < 0.05$), social functioning ($p < 0.05$ and 'physical health summary' ($p = 0.02$) on SF-36 survey.		Low (single study, very small sample size)
BMP4	Wang et al. (2013) [25]	6007C>T T allele: OR 0.51, $p < 0.001$		6007C>T T allele: OR of improvement 1.53, $p = 0.002$	Low (inconsistency across studies, inconsistency between CSM and OPLL studies)
COL9A2	Wang et al. (2012) [30]	Trp2+ allele: OR 1.78, $p = 0.048$			Low (single study, small sample size)
COL11A2	Koga et al. (1998) [31]	Promoter ($-$182) C allele ($p = 0.0240$); Intron 6 ($-$4) T allele ($p = 0.0004$); Exon 43 (+24) G allele ($p = 0.0210$); Exon 46 (+18) T allele ($p = 0.0333$)			Low
HIF1A	Wang et al. (2014) [35]	1790G>A A allele: OR 1.62, $p < 0.001$	1790G>A A allele associated with worse mJOA scores ($p < 0.001$)	1790G>A A allele: OR of improvement 1.55, $p = 0.024$	Low (single study)
IL15RA	Guo et al. (2014) [37]	rs2228059 A allele: OR 1.63, $p < 0.001$			Low
NPPS	Nakamura et al. (1999) [39]	IVS20-11delT: $p = 0.0029$			Low (inconsistency across studies)
	Koshizuka et al. (2002) [40]	IVS15-14T>C: OR 3.01, $p = 0.022$ NB. Horikoshi et al. (2006) find $p = 0.320$.			
OPG	Yu et al. (2018) [42]	950T>C C allele: $p < 0.01$	950T>C TT genotype associated with higher mJOA scores and fewer ossified vertebrae ($p < 0.05$)		Low (single study)
OPN	Wu et al. (2014) [43]	$-$66T>G G allele: OR 1.55, $p = 0.002$	No significant difference in mJOA score ($p > 0.05$).	$-$66T>G GG genotype: OR of no improvement 3.62, $p = 0.007$	Low (single study)
RUNX2	Chang et al. (2017) [44]	rs967588C>T T allele: OR 0.47, $p = 0.033$; rs16873379T>C C allele: OR 0.48, $p = 0.033$; rs1406846T>A A allele: OR 5.67, $p < 0.001$		rs16873379T>C C allele: $p < 0.05$; rs6908650G>A A allele: $p < 0.05$; rs1406846T>A A allele: $p < 0.05$; rs2677108T>C C allele: $p < 0.05$	Low
TGFB3	Horikoshi et al. (2006) [33]	IVS1-1284G>C CC genotype: OR 1.46, $p = 0.044$			Low (single study)
VDR	Wang et al. (2010) [51]	*ApaI* A allele: OR 2.88, $p < 0.001$; *TaqI* C allele: OR 4.67, $p < 0.001$	No significant difference in mJOA score ($p > 0.05$).		Low (inconsistency across studies)
	Song et al. (2018) [49]	*FokI* ff genotype: OR 1.985, $p = 0.003$	No significant difference in mJOA score or number of ossified vertebrae ($p > 0.05$)		

4.1. Spinal Column Disease: Focus on OPLL

The greatest focus of research to date has been on the bone morphogenetic proteins, a group of multifunctional growth factors that fall within the *TGFβ* superfamily and are involved in cartilage development and the induction of bone formation [55]. Four genes within this family of growth factors have been associated with both altered susceptibilities to bony spinal pathology and altered susceptibility to the development of myelopathy: *BMP2*, *BMP4*, *BMP9*, and *BMPR1A*. The 4A>C SNP in the *BMPR1A* gene is associated with a significantly greater likelihood of radiological OPLL and a significantly greater number of ossified vertebrae [27]. Similarly, the CTCA haplotype of the *BMP9* gene is associated with a significantly increased risk of developing OPLL (OR 2.37), as well as a greater number of ossified vertebrae [26]. In the *BMP4* gene, a haplotype of 7 SNPs is associated with both greater susceptibilities to OPLL and worse disease [24]. Moreover, the 6007C>T SNP in the *BMP4* gene is associated with not only greater likelihood of developing bony pathology and greater severity of radiological disease, but also a greater likelihood of post-operative improvement of the mJOA score [23,25].

The dual role of 6007C>T SNP in the *BMP4* gene merits further discussion. The T allele of the polymorphism was found to be protective for spinal cord disease [25] (AOR 0.51) and it was associated with better outcomes in mJOA score after surgery (AOR 1.53 of being in the 'improvement' group). Conversely, Meng et al. found the same T allele to be associated with a greater likelihood of radiological OPLL (OR 1.57) [23]. The contrasting effect of the same allele suggests the effect of the *BMP4* gene is not limited to spinal pathology and the development of bony compression, but it may also influence the spinal cord response to such compression. It is unclear whether this effect is due to an intrinsic effect of *BMP4* on CNS resilience or regeneration, or a treatment artifact that faster compression elicited by the 6007C>T polymorphism giving more severe bony pathology results in faster decompression and better post-operative outcomes. Nonetheless, it is clear that bone morphogenetic protein genes may have extensive influences in the pathogenesis and symptoms of DCM.

Alongside the *BMP* genes, several other genes should be highlighted. In the *NPPS* gene, the C973T polymorphism significantly affected both the susceptibility of OPLL development and the thickness of ossified vertebrae, but notably did not affect the number of ossified vertebrae.

NPPS gene polymorphisms were implicated in post-surgical improvements of spinal column disease affecting the thickness of ossified vertebrae (C973T), while others (IVS15-14T>C) affect the number of ossified vertebrae and others affect both (IVS20-11delT) [41].

Evaluation of the network of genes that were found to be associated with the development of spinal column pathology shows that, while each gene has an independent effect on susceptibility to pathology, there is clear connectedness within and across gene families (Figure 3).

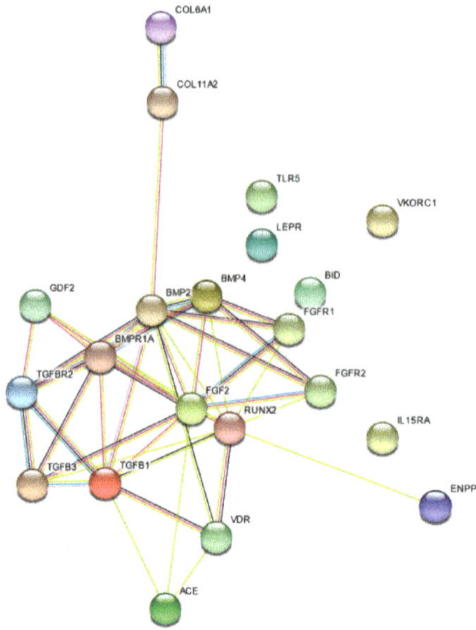

Figure 3. STRING Evidence Network for genes associated with spinal column disease.

4.2. Spinal Cord Disease

The ε4 allele of the *APOE* gene, an allele that is well known for its associations with both cardiovascular disease and Alzheimer's disease, was associated with both a significantly increased likelihood of DCM development (OR 3.50) [16] and a significantly greater likelihood of failing to gain post-operative improvement (AOR 8.60 no improvement) [54]. However, this effect might not be universal across ethnicities; a study in an Indian population found the ε2 allele to be associated with development of myelopathy (OR 6.69) [17].

The 1790G>A polymorphism of the *HIF1A* gene displayed the opposite effect: it was associated with significantly greater likelihood of DCM development (OR 1.62), and worse disease but a greater likelihood of post-surgical improvement (OR 1.55) [35].

Reductions in Hif1α expression have been shown to be associated with the neuroprotective benefits of hyperbaric oxygen in spinal cord injury mouse models [56]. It is possible that such a mechanism is also the mediator of the *HIF1A* polymorphism's effect on susceptibility, severity, and post-operative response in DCM.

The *APOE* gene and its product, the apolipoprotein E transporter, are well-known to be involved remyelination, with defective clearance of myelin debris by the transporter limiting the potential for remyelination [57]. In the case of both *HIF1A* and *APOE*, their effects appear to be directly exerted on the cord's response to bony pathology, rather than via the bony pathology itself.

There appears to be delineation between genetic factors contributing to the development of bony pathology in the cervical spine, and those contributing to the CNS response to such insult. That an SNP of brain-derived neurotrophic factor (*BDNF*) is associated with the severity of disability (i.e., CNS response to insult) gives further weight to such a distinction [53].

As with genes that are associated with spinal pathology, the genes studied with relation to spinal cord disease have independent, but connected, effects (Figure 4).

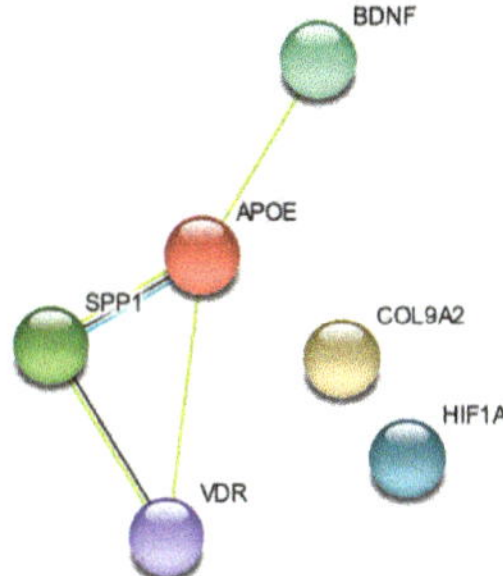

Figure 4. STRING Evidence Network diagram for genes associated with spinal cord disease.

4.3. Conflicting Evidence

The frequency of conflicting evidence is one striking aspect of much of the work reviewed here. The best example of this is perhaps seen in the *RUNX2* gene; the rs1406846 SNP A allele is associated with 5.67 times greater likelihood of developing DCM in one study [44], but it has no significant effect in a further study using a similar number of participants from the same country [20]. Similarly the 869T>C SNP in the *TGFB1* gene was associated with an odds ratio of 4.50 in one study [45], but a larger, more recent study found no significant effect of the same allele [46], with the result of meta-analysis showing no significant effect. Further examples of conflicting evidence include the IVS20-11delT polymorphism of the *NPPS* gene, in which one study found a significant effect on DCM susceptibility [39], but two others found no significant effect [38,41], while in the IVS15-14T>C polymorphism, two studies found an effect on susceptibility [40,41], with a further study showing no significant effect [33]. Such inconsistency might reflect the relatively small sample sizes of much of the work described here, and it indicates the need for large, well powered genetic investigations.

4.4. Limitations of Current Work

Limitations of the current work on the genetics of DCM are multiple. Firstly, many of the studies that were reviewed in this article scored poorly on the MINORS methodological items assessment [13]. None published information regarding prospective calculation of study size, few reported whether the cases and controls were demographically matched, and some did not report how participants were recruited (e.g., consecutively). As mentioned above, the sample sizes remain in the hundreds rather than thousands, which limits the degree to which their conclusions can be considered valid. Moreover, in reporting the results, many omit odds ratios, instead of reporting only *p*-values, which limits the degree to which such results can be interpreted.

Many of the studies reviewed here focused exclusively on Japanese, Chinese, or South Korean participants, and specifically OPLL. Interestingly, in the *APOE* gene ethnicity appears to result in conflicting genetic effects, with the ε2 allele associated with myelopathy in Indian populations and the ε4 allele associated with myelopathy in Chinese populations [16,17]. It is widely acknowledged that there is a greater prevalence of OPLL within Asian populations, and this might explain their disproportionate representation in the literature [1]. However, without further work across ethnicities, it remains speculation as to whether the conclusions from these studies are globally relevant and across the spectrum of DCM pathologies.

There is significant diversity in the assessment of disease severity between studies. One study used the SF-36 quality of life survey [53], three used the mJOA score [35,43,51] (a clinical score commonly used in DCM research [58–61]), while others used radiographic measures [19,23,24,26,27,41,44,51]. A similar situation is found within the literature while considering response to surgery, with one study using a cut-off for 'improvement' as +1 point on mJOA score [54], some using >50% increase in mJOA score [25,35,43], one using a t-test of % improvement on mJOA between homozygous groups [44], and one paper while using a radiographic definition of disease progression [41]. Such heterogeneity of outcome measures limits the degree to which the effects of genes on severity of DCM and response to surgery can be compared. The removal of surrogate outcome measures and more consistent use of a single form of outcome measure would permit more readily comparable conclusions to be drawn across different studies. We are currently undertaking RECODE DCM, an international consensus process to standardize the reporting of data elements in DCM research, and this would clearly hold benefit here (www.recode-dcm.com) [62]. For the reasons that are outlined above, the GRADE ratings of quality of evidence for each candidate gene were 'low' across all genes.

4.5. Future Directions

It is clear that interest in this field is building, with increasing numbers of studies focusing on genetic effects in DCM (Figure 5). However, more than half the that are genes reviewed here have been investigated by only a single study, often with small sample sizes, which suggests more intensive work in larger populations is required to further describe the genetic basis of DCM. Furthermore, all of the studies included in this review focused on individual candidate genes. While some considered the effects of haplotypes consisting of several SNPs within a single gene [24,26,29], no work has yet combined SNPs across different genes. Such combinations may exhibit effect sizes of greater magnitude than those in the current body of literature, with potential for such genetic profiles permitting greater personalization of treatment strategies. Future work should also seek to characterize the mechanism by which the genes that were reviewed here exert their effects in the pathobiology of DCM.

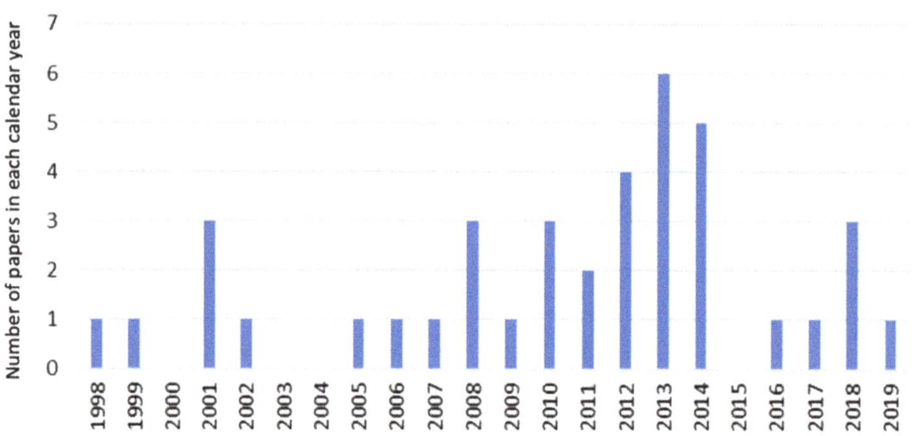

Figure 5. Bar graph of number of papers investigating candidate genes in DCM in each calendar year.

5. Conclusions

While a number of limitations of the current work do exist, there is clear evidence of genetic effects of single nucleotide polymorphisms and haplotypes in DCM. Some of the genes exert their influence on the development of bony pathology, while others have effects on the spinal cord itself. Further investigation of the genetic basis of DCM requires larger study sizes, using more consistent measures of disease severity and response to surgery. The current evidence base is insufficient for translation to clinical practice for use in prognostication and management, but the potential for genetic profiles to be used in this way may well be realized once greater characterization of the genetic basis of DCM is achieved.

Supplementary Materials: The following are available online at http://www.mdpi.com/2077-0383/9/1/282/s1, Data S1: search strategy used for MEDLINE and EMBASE databases. Data S2: search strategy used for Cochrane and HuGENet databases. Data S3: PRSIMA checklist.

Author Contributions: Conceptualization, B.M.D. and M.R.N.K.; methodology, B.M.D. and D.H.P.; formal analysis B.M.D. and D.H.P.; writing—original draft preparation, D.H.P.; writing—review and editing, B.M.D., D.H.P., O.D.M., A.R.B. and M.R.N.K.; supervision, M.R.N.K. All authors have read and agreed to the published version of the manuscript.

Funding: Research in the senior author's laboratory is supported by a core support grant from the Wellcome Trust and MRC to the Wellcome Trust-Medical Research Council Cambridge Stem Cell Institute. MRNK is supported by a NIHR Clinician Scientist Award.

Acknowledgments: The authors are grateful to the Medical Library at the University of Cambridge for their assistance in the design of the search strategy for this article.

Conflicts of Interest: The authors declare no conflict of interest. The funders had no role in the design of the study; in the collection, analyses, or interpretation of data; in the writing of the manuscript, or in the decision to publish the results.

References

1. Nouri, A.; Tetreault, L.; Singh, A.; Karadimas, S.K.; Fehlings, M.G. Degenerative Cervical Myelopathy: Epidemiology, Genetics, and Pathogenesis. *Spine* **2015**, *40*, E675–E693. [CrossRef] [PubMed]

2. Davies, B.M.; Mowforth, O.D.; Smith, E.K.; Kotter, M.R. Degenerative cervical myelopathy. *BMJ* **2018**, *360*, k186. [CrossRef] [PubMed]

3. Tetreault, L.; Goldstein, C.L.; Arnold, P.; Harrop, J.; Hilibrand, A.; Nouri, A.; Fehlings, M.G. Degenerative Cervical Myelopathy: A Spectrum of Related Disorders Affecting the Aging Spine. *Neurosurgery* **2015**, *77*, S51–S67. [CrossRef] [PubMed]

4. Kalsi-Ryan, S.; Karadimas, S.K.; Fehlings, M.G. Cervical Spondylotic Myelopathy: The Clinical Phenomenon and the Current Pathobiology of an Increasingly Prevalent and Devastating Disorder. *Neuroscientist* **2013**, *19*, 409–421. [CrossRef]

5. Nouri, A.; Tetreault, L.; Côté, P.; Zamorano, J.J.; Dalzell, K.; Fehlings, M.G. Does Magnetic Resonance Imaging Improve the Predictive Performance of a Validated Clinical Prediction Rule Developed to Evaluate Surgical Outcome in Patients with Degenerative Cervical Myelopathy? *Spine* **2015**, *40*, 1092–1100. [CrossRef]

6. Witiw, C.D.; Mathieu, F.; Nouri, A.; Fehlings, M.G. Clinico-Radiographic Discordance: An Evidence-Based Commentary on the Management of Degenerative Cervical Spinal Cord Compression in the Absence of Symptoms or With Only Mild Symptoms of Myelopathy. *Glob. Spine J.* **2018**, *8*, 527–534. [CrossRef]

7. Tempest-Mitchell, J.; Hilton, B.; Davies, B.M.; Nouri, A.; Hutchinson, P.J.; Scoffings, D.J.; Mannion, R.J.; Trivedi, R.; Timofeev, I.; Crawford, J.R.; et al. A comparison of radiological descriptions of spinal cord compression with quantitative measures, and their role in non-specialist clinical management. *PLoS ONE* **2019**, *14*, e021938. [CrossRef]

8. Yoon, S.T.; Raich, A.; Hashimoto, R.E.; Riew, K.D.; Shaffrey, C.I.; Rhee, J.M.; Tetreault, L.A.; Skelly, A.C.; Fehlings, M.G. Predictive factors affecting outcome after cervical laminoplasty. *Spine* **2013**, *38*, S232–S252. [CrossRef]

9. Patel, A.A.; Spiker, W.R.; Daubs, M.; Brodke, D.S.; Cannon-Albright, L.A. Evidence of an Inherited Predisposition for Cervical Spondylotic Myelopathy. *Spine* **2012**, *37*, 26–29. [CrossRef]

10. Wilson, J.R.; Patel, A.A.; Brodt, E.D.; Dettori, J.R.; Brodke, D.S.; Fehlings, M.G. Genetics and Heritability of Cervical Spondylotic Myelopathy and Ossification of the Posterior Longitudinal Ligament: Results of a Systematic Review. *Spine* **2013**, *38*, S123–S146. [CrossRef]

11. Wang, G.; Cao, Y.; Wu, T.; Duan, C.; Wu, J.; Hu, J.; Lu, H. Genetic factors of cervical spondylotic myelopathy-a systemic review. *J. Clin. Neurosci.* **2017**, *44*, 89–94. [CrossRef] [PubMed]

12. Moher, D.; Liberati, A.; Tetzlaff, J.; Altman, D.G.; Group, T.P. Preferred Reporting Items for Systematic Reviews and Meta-Analyses: The PRISMA Statement. *PLoS Med.* **2009**, *6*, e1000097. [CrossRef] [PubMed]

13. Slim, K.; Nini, E.; Forestier, D.; Kwiatkowski, F.; Panis, Y.; Chipponi, J. Methodological index for non-randomized studies (MINORS): Development and validation of a new instrument. *ANZ J. Surg.* **2003**, *73*, 712–716. [CrossRef] [PubMed]

14. Balshem, H.; Helfand, M.; Schünemann, H.J.; Oxman, A.D.; Kunz, R.; Brozek, J.; Vist, G.E.; Falck-Ytter, Y.; Meerpohl, J.; Norris, S.; et al. GRADE guidelines: 3. Rating the quality of evidence. *J. Clin. Epidemiol.* **2011**, *64*, 401–406. [CrossRef]

15. Kim, D.H.; Yun, D.H.; Kim, H.-S.; Min, S.K.; Yoo, S.D.; Lee, K.H.; Kim, K.-T.; Jo, D.J.; Kim, S.K.; Chung, J.-H.; et al. The Insertion/Deletion Polymorphism of Angiotensin I Converting Enzyme Gene is Associated with Ossification of the Posterior Longitudinal Ligament in the Korean Population. *Ann. Rehabil. Med.* **2014**, *38*, 1–5. [CrossRef]

16. Setzer, M.; Hermann, E.; Seifert, V.; Marquardt, G. Apolipoprotein E gene polymorphism and the risk of cervical myelopathy in patients with chronic spinal cord compression. *Spine* **2008**, *33*, 497–502. [CrossRef]

17. Diptiranjan, S.; Harshitha, S.M.; Sibin, M.K.; Arati, S.; Chetan, G.K.; Bhat, D.I. Role of APOE and IL18RAP gene polymorphisms in cervical spondylotic myelopathy in Indian population. *J. Clin. Neurosci.* **2019**, *66*, 83–86. [CrossRef]

18. Chon, J.; Hong, J.-H.; Kim, J.; Han, Y.J.; Lee, B.W.; Kim, S.-C.; Kim, D.H.; Yoo, S.D.; Kim, H.-S.; Yun, D.H. Association between BH3 interacting domain death agonist (BID) gene polymorphism and ossification of the posterior longitudinal ligament in Korean population. *Mol. Biol. Rep.* **2014**, *41*, 895–899. [CrossRef]

19. Wang, H.; Liu, D.; Yang, Z.; Tian, B.; Li, J.; Meng, X.; Wang, Z.; Yang, H.; Lin, X. Association of bone morphogenetic protein-2 gene polymorphisms with susceptibility to ossification of the posterior longitudinal ligament of the spine and its severity in Chinese patients. *Eur. Spine J.* **2008**, *17*, 956–964. [CrossRef]

20. Liu, Y.; Zhao, Y.; Chen, Y.; Shi, G.; Yuan, W. RUNX2 Polymorphisms Associated with OPLL and OLF in the Han Population. *Clin. Orthop. Relat. Res.* **2010**, *468*, 3333–3341. [CrossRef]

21. Yan, L.; Chang, Z.; Liu, Y.; Li, Y.-B.; He, B.-R.; Hao, D.-J. A single nucleotide polymorphism in the human bone morphogenetic protein-2 gene (109T>G) affects the Smad signaling pathway and the predisposition to ossification of the posterior longitudinal ligament of the spine. *Chin. Med. J.* **2013**, *126*, 1112–1118.

22. Kim, K.H.; Kuh, S.U.; Park, J.Y.; Lee, S.J.; Park, H.S.; Chin, D.K.; Kim, K.S.; Cho, Y.E. Association between BMP-2 and COL6A1 gene polymorphisms with susceptibility to ossification of the posterior longitudinal ligament of the cervical spine in Korean patients and family members. *Genet. Mol. Res.* **2014**, *13*, 2240–2247. [CrossRef] [PubMed]

23. Meng, X.; Wang, H.; Yang, H.; Hai, Y.; Tian, B.; Lin, X. T allele at site 6007 of bone morphogenetic protein-4 gene increases genetic susceptibility to ossification of the posterior longitudinal ligament in male Chinese Han population. *Chin. Med. J.* **2010**, *123*, 2537–2542. [PubMed]

24. Ren, Y.; Feng, J.; Liu, Z.; Wan, H.; Li, J.; Lin, X. A new haplotype in BMP4 implicated in ossification of the posterior longitudinal ligament (OPLL) in a Chinese population. *J. Orthop. Res.* **2012**, *30*, 748–756. [CrossRef]

25. Wang, D.; Liu, W.; Cao, Y.; Yang, L.; Liu, B.; Yao, G.; Bi, Z. BMP-4 polymorphisms in the susceptibility of cervical spondylotic myelopathy and its outcome after anterior cervical corpectomy and fusion. *Cell. Physiol. Biochem.* **2013**, *32*, 210–217. [CrossRef] [PubMed]

26. Ren, Y.; Liu, Z.; Feng, J.; Wan, H.; Li, J.; Wang, H.; Lin, X. Association of a BMP9 Haplotype with Ossification of the Posterior Longitudinal Ligament (OPLL) in a Chinese Population. *PLoS ONE* **2012**, *7*, e40587. [CrossRef] [PubMed]

27. Wang, H.; Jin, W.; Li, H. Genetic polymorphisms in bone morphogenetic protein receptor type IA gene predisposes individuals to ossification of the posterior longitudinal ligament of the cervical spine via the smad signaling pathway. *BMC Musculoskelet. Disord.* **2018**, *19*, 61. [CrossRef] [PubMed]

28. Tanaka, T.; Ikari, K.; Furushima, K.; Okada, A.; Tanaka, H.; Furukawa, K.-I.; Yoshida, K.; Ikeda, T.; Ikegawa, S.; Hunt, S.C.; et al. Genomewide Linkage and Linkage Disequilibrium Analyses Identify COL6A1, on Chromosome 21, as the Locus for Ossification of the Posterior Longitudinal Ligament of the Spine. *Am. J. Hum. Genet.* **2003**, *73*, 812–822. [CrossRef]

29. Kong, Q.; Ma, X.; Li, F.; Guo, Z.; Qi, Q.; Li, W.; Yuan, H.; Wang, Z.; Chen, Z. COL6A1 Polymorphisms Associated with Ossification of the Ligamentum Flavum and Ossification of the Posterior Longitudinal Ligament. *Spine* **2007**, *32*, 2834–2838. [CrossRef]

30. Wang, Z.C.; Shi, J.G.; Chen, X.S.; Xu, G.H.; Li, L.J.; Jia, L.S. The role of smoking status and collagen IX polymorphisms in the susceptibility to cervical spondylotic myelopathy. *Genet. Mol. Res.* **2012**, *11*, 1238–1244. [CrossRef]

31. Koga, H.; Sakou, T.; Taketomi, E.; Hayashi, K.; Numasawa, T.; Harata, S.; Yone, K.; Matsunaga, S.; Otterud, B.; Inoue, I.; et al. Genetic mapping of ossification of the posterior longitudinal ligament of the spine. *Am. J. Hum. Genet.* **1998**, *62*, 1460–1467. [CrossRef] [PubMed]

32. Maeda, S.; Ishidou, Y.; Koga, H.; Taketomi, E.; Ikari, K.; Komiya, S.; Takeda, J.; Sakou, T.; Inoue, I. Functional Impact of Human Collagen α2(XI) Gene Polymorphism in Pathogenesis of Ossification of the Posterior Longitudinal Ligament of the Spine. *J. Bone Miner. Res.* **2001**, *16*, 948–957. [CrossRef] [PubMed]

33. Horikoshi, T.; Maeda, K.; Kawaguchi, Y.; Chiba, K.; Mori, K.; Koshizuka, Y.; Hirabayashi, S.; Sugimori, K.; Matsumoto, M.; Kawaguchi, H.; et al. A large-scale genetic association study of ossification of the posterior longitudinal ligament of the spine. *Hum. Genet.* **2006**, *119*, 611–616. [CrossRef] [PubMed]

34. Jun, J.-K.; Kim, S.-M. Association Study of Fibroblast Growth Factor 2 and Fibroblast Growth Factor Receptors Gene Polymorphism in Korean Ossification of the Posterior Longitudinal Ligament Patients. *J. Korean Neurosurg. Soc.* **2012**, *52*, 7–13. [CrossRef]

35. Wang, Z.-C.; Hou, X.-W.; Shao, J.; Ji, Y.-J.; Li, L.; Zhou, Q.; Yu, S.-M.; Mao, Y.-L.; Zhang, H.-J.; Zhang, P.-C.; et al. HIF-1α polymorphism in the susceptibility of cervical spondylotic myelopathy and its outcome after anterior cervical corpectomy and fusion treatment. *PLoS ONE* **2014**, *9*, e110862. [CrossRef]

36. Kim, D.H.; Jeong, Y.S.; Chon, J.; Yoo, S.D.; Kim, H.-S.; Kang, S.W.; Chung, J.-H.; Kim, K.-T.; Yun, D.H. Association between interleukin 15 receptor, alpha (IL15RA) polymorphism and Korean patients with ossification of the posterior longitudinal ligament. *Cytokine* **2011**, *55*, 343–346. [CrossRef]

37. Guo, Q.; Lv, S.-Z.; Wu, S.-W.; Tian, X.; Li, Z.-Y. Association between single nucleotide polymorphism of IL15RA gene with susceptibility to ossification of the posterior longitudinal ligament of the spine. *J. Orthop. Surg. Res.* **2014**, *9*, 103. [CrossRef]

38. Tahara, M.; Aiba, A.; Yamazaki, M.; Ikeda, Y.; Goto, S.; Moriya, H.; Okawa, A. The extent of ossification of posterior longitudinal ligament of the spine associated with nucleotide pyrophosphatase gene and leptin receptor gene polymorphisms. *Spine* **2005**, *30*, 877–880. [CrossRef]

39. Nakamura, I.; Ikegawa, S.; Okawa, A.; Okuda, S.; Koshizuka, Y.; Kawaguchi, H.; Nakamura, K.; Koyama, T.; Goto, S.; Toguchida, J.; et al. Association of the human NPPS gene with ossification of the posterior longitudinal ligament of the spine (OPLL). *Hum. Genet.* **1999**, *104*, 492–497. [CrossRef]

40. Koshizuka, Y.; Kawaguchi, H.; Ogata, N.; Ikeda, T.; Mabuchi, A.; Seichi, A.; Nakamura, Y.; Nakamura, K.; Ikegawa, S. Nucleotide Pyrophosphatase Gene Polymorphism Associated with Ossification of the Posterior Longitudinal Ligament of the Spine. *J. Bone Miner. Res.* **2002**, *17*, 138–144. [CrossRef]

41. He, Z.; Zhu, H.; Ding, L.; Xiao, H.; Chen, D.; Xue, F. Association of NPP1 polymorphism with postoperative progression of ossification of the posterior longitudinal ligament in Chinese patients. *Genet. Mol. Res.* **2013**, *12*, 4648–4655. [CrossRef] [PubMed]

42. Yu, H.-M.; Chen, X.-L.; Wei, W.; Yao, X.-D.; Sun, J.-Q.; Su, X.-T.; Lin, S.-F. Effect of osteoprotegerin gene polymorphisms on the risk of cervical spondylotic myelopathy in a Chinese population. *Clin. Neurol. Neurosurg.* **2018**, *175*, 149–154. [CrossRef] [PubMed]

43. Wu, J.; Wu, D.; Guo, K.; Yuan, F.; Ran, B. OPN Polymorphism is Associated with the Susceptibility to Cervical Spondylotic Myelopathy and its Outcome After Anterior Cervical Corpectomy and Fusion. *CPB* **2014**, *34*, 565–574. [CrossRef]

44. Chang, F.; Li, L.; Gao, G.; Ding, S.; Yang, J.; Zhang, T.; Zuo, G. Role of Runx2 polymorphisms in risk and prognosis of ossification of posterior longitudinal ligament. *J. Clin. Lab. Anal.* **2017**, *31*, e22068. [CrossRef] [PubMed]

45. Kamiya, M.; Harada, A.; Mizuno, M.; Iwata, H.; Yamada, Y. Association between a polymorphism of the transforming growth factor-beta1 gene and genetic susceptibility to ossification of the posterior longitudinal ligament in Japanese patients. *Spine* **2001**, *26*, 1264–1266. [CrossRef] [PubMed]

46. Han, I.B.; Ropper, A.E.; Jeon, Y.J.; Park, H.S.; Shin, D.A.; Teng, Y.D.; Kuh, S.-U.; Kim, N.-K. Association of transforming growth factor-beta 1 gene polymorphism with genetic susceptibility to ossification of the posterior longitudinal ligament in Korean patients. *Genet. Mol. Res.* **2013**, *12*, 4807–4816. [CrossRef]

47. Jekarl, D.W.; Paek, C.-M.; An, Y.J.; Kim, Y.J.; Kim, M.; Kim, Y.; Lee, J.; Sung, C.H. TGFBR2 gene polymorphism is associated with ossification of the posterior longitudinal ligament. *J. Clin. Neurosci.* **2013**, *20*, 453–456. [CrossRef]

48. Chung, W.-S.; Nam, D.-H.; Jo, D.-J.; Lee, J.-H. Association of Toll-Like Receptor 5 Gene Polymorphism with Susceptibility to Ossification of the Posterior Longitudinal Ligament of the Spine in Korean Population. *J. Korean Neurosurg. Soc.* **2011**, *49*, 8–12. [CrossRef]

49. Song, D.; Wu, Y.; Tian, D. Association of VDR-FokI and VDBP-Thr420Lys polymorphisms with cervical spondylotic myelopathy: A case-control study in the population of China. *J. Clin. Lab. Anal.* **2018**, *33*, e22669. [CrossRef]

50. Kobashi, G.; Ohta, K.; Washio, M.; Okamoto, K.; Sasaki, S.; Yokoyama, T.; Miyake, Y.; Sakamoto, N.; Hata, A.; Tamashiro, H.; et al. FokI Variant of Vitamin D Receptor Gene and Factors Related to Atherosclerosis Associated With Ossification of the Posterior Longitudinal Ligament of the Spine: A Multi-Hospital Case-Control Study. *Spine* **2008**, *33*, E553–E558. [CrossRef]

51. Wang, Z.C.; Chen, X.S.; Wang, D.W.; Shi, J.G.; Jia, L.S.; Xu, G.H.; Huang, J.H.; Fan, L. The genetic association of Vitamin D receptor polymorphisms and cervical spondylotic myelopathy in Chinese subjects. *Clin. Chim. Acta* **2010**, *411*, 794–797. [CrossRef] [PubMed]

52. Chin, D.-K.; Han, I.-B.; Ropper, A.E.; Jeon, Y.-J.; Kim, D.-H.; Kim, Y.-S.; Park, Y.; Teng, Y.D.; Kim, N.-K.; Kuh, S.-U. Association of VKORC1-1639G> A polymorphism with susceptibility to ossification of the posterior longitudinal ligament of the spine: A Korean study. *Acta Neurochir.* **2013**, *155*, 1937–1942. [CrossRef] [PubMed]

53. Abode-Iyamah, K.O.; Stoner, K.E.; Grossbach, A.J.; Viljoen, S.V.; McHenry, C.L.; Petrie, M.A.; Dahdaleh, N.S.; Grosland, N.M.; Shields, R.K.; Howard, M.A.; et al. Effects of brain derived neurotrophic factor Val66Met polymorphism in patients with cervical spondylotic myelopathy. *J. Clin. Neurosci.* **2016**, *24*, 117–121. [CrossRef] [PubMed]

54. Setzer, M.; Vrionis, F.D.; Hermann, E.J.; Seifert, V.; Marquardt, G. Effect of apolipoprotein E genotype on the outcome after anterior cervical decompression and fusion in patients with cervical spondylotic myelopathy. *J. Neurosurg. Spine* **2009**, *11*, 659–666. [CrossRef]

55. Chen, D.; Zhao, M.; Mundy, G.R. Bone morphogenetic proteins. *Growth Factors* **2004**, *22*, 233–241. [CrossRef]

56. Zhou, Y.; Liu, X.; Qu, S.; Yang, J.; Wang, Z.; Gao, C.; Su, Q. Hyperbaric oxygen intervention on expression of hypoxia-inducible factor-1α and vascular endothelial growth factor in spinal cord injury models in rats. *Chin. Med. J.* **2013**, *126*, 3897–3903.

57. Cantuti-Castelvetri, L.; Fitzner, D.; Bosch-Queralt, M.; Weil, M.-T.; Su, M.; Sen, P.; Ruhwedel, T.; Mitkovski, M.; Trendelenburg, G.; Lütjohann, D.; et al. Defective cholesterol clearance limits remyelination in the aged central nervous system. *Science* **2018**, *359*, 684–688. [CrossRef]

58. Kalsi-Ryan, S.; Singh, A.R.; Massicotte, E.M.M.; Arnold, P.M.M.; Brodke, D.S.M.; Norvell, D.C.; Hermsmeyer, J.T.; Fehlings, M.G. Ancillary Outcome Measures for Assessment of Individuals with Cervical Spondylotic Myelopathy. *Spine* **2013**, *38*, S111–S122. [CrossRef]

59. Davies, B.M.; McHugh, M.; Elgheriani, A.; Kolias, A.G.; Tetreault, L.A.; Hutchinson, P.J.A.; Fehlings, M.G.; Kotter, M.R.N. Reported Outcome Measures in Degenerative Cervical Myelopathy: A Systematic Review. *PLoS ONE* **2016**, *11*, e0157263. [CrossRef]

60. Davies, B.M.; McHugh, M.; Elgheriani, A.; Kolias, A.G.; Tetreault, L.; Hutchinson, P.J.A.; Fehlings, M.G.; Kotter, M.R.N. The reporting of study and population characteristics in degenerative cervical myelopathy: A systematic review. *PLoS ONE* **2017**, *12*, e0172564. [CrossRef]

61. Kopjar, B.; Tetreault, L.; Kalsi-Ryan, S.; Fehlings, M. Psychometric properties of the modified japanese Orthopaedic association scale in patients with cervical spondylotic myelopathy. *Spine* **2015**, *40*, E23–E28. [CrossRef] [PubMed]

62. Davies, B.M.; Khan, D.Z.; Mowforth, O.D.; McNair, A.G.K.; Gronlund, T.; Kolias, A.G.; Tetreault, L.; Starkey, M.L.; Sadler, I.; Sarewitz, E.; et al. RE-CODE DCM (REsearch Objectives and Common Data Elements for Degenerative Cervical Myelopathy): A Consensus Process to Improve Research Efficiency in DCM, Through Establishment of a Standardized Dataset for Clinical Research and the Definition of the Research Priorities. *Glob. Spine J.* **2019**, *9*, 65S–76S.

Journal of
Clinical Medicine

Review

Snake-Eye Myelopathy and Surgical Prognosis: Case Series and Systematic Literature Review

Marco Maria Fontanella [1,*], Luca Zanin [1], Riccardo Bergomi [1], Marco Fazio [2], Costanza Maria Zattra [1], Edoardo Agosti [3], Giorgio Saraceno [1], Silvia Schembari [3], Lucio De Maria [1], Luisa Quartini [4], Ugo Leggio [5], Massimiliano Filosto [6], Roberto Gasparotti [7] and Davide Locatelli [3]

[1] Neurosurgery Unit, Department of Medical and Surgical Specialties, Radiological Sciences and Public Health, University of Brescia, 25123 Brescia, Italy; lucazanin00@gmail.com (L.Z.); rbergomi@tiscalinet.it (R.B.); costanzamaria.zattra01@universitadipavia.it (C.M.Z.); g.saraceno@unibs.it (G.S.); luciodemaria@gmail.com (L.D.M.)

[2] Neurosurgery Unit, Poliambulanza Foundation, 24124 Brescia, Italy; marcofaziomd@gmail.com

[3] Neurosurgery Unit, Department of Biotechnology and Life Sciences (DBSV), University of Insubria, Ospedale di Circolo e Fondazione Macchi, 21100 Varese, Italy; edoardo_agosti@libero.it (E.A.); silvia.schembari@outlook.com (S.S.); davide1.locatelli@uninsubria.it (D.L.)

[4] Intensive Care Unit, Department of Anesthesia, Intensive Care and Emergency, ASST Spedali Civili di Brecia, 25123 Brescia, Italy; neuro.chirurgiabs@gmail.com

[5] Neurophysiopathology Unit, Department of Neurological Sciences and Vision, ASST Spedali Civili di Brecia, 25123 Brescia, Italy; ugo.leggio@unibs.it

[6] Center for Neuromuscular Diseases, Unit of Neurology, ASST "Spedali Civili", 25123 Brescia, Italy; massimiliano.filosto@unibs.it

[7] Neuroradiology Unit, Department of Medical and Surgical Specialties, Radiological Sciences, and Public Health, University of Brescia, 25123 Brescia, Italy; roberto.gasparotti@unibs.it

* Correspondence: marco.fontanella@unibs.it; Tel.: +39-030-3995-587

Received: 9 May 2020; Accepted: 9 July 2020; Published: 12 July 2020

Abstract: The prognostic value of "snake-eyes" sign in spinal cord magnetic resonance imaging (MRI) is unclear and the correlation with different pathological conditions has not been completely elucidated. In addition, its influence on surgical outcome has not been investigated in depth. A literature review according to PRISMA (Preferred reporting items for systematic review and meta-analysis protocols) guidelines on the prognostic significance of "snake-eyes" sign in operated patients was performed. Clinical, neuroradiological, and surgical data of three institutional patients, were also retrospectively collected. The three patients, with radiological evidence of "snake-eyes" myelopathy, underwent appropriate surgical treatment for their condition, with no new post-operative neurological deficits and good outcome at follow-up. The literature review, however, reported conflicting results: the presence of "snake-eyes" sign seems a poor prognostic factor in degenerative cervical myelopathy, even if some cases can improve after surgery. "Snake-eyes" myelopathy represents a rare form of myelopathy; pathophysiology is still unclear. The frequency of this myelopathy may be greater than previously thought and according to our literature review it is mostly a negative prognostic factor. However, from our experience, prognosis might not be so dire, especially when tailored surgical intervention is performed; therefore, surgery should always be considered and based on the complete clinical, neurophysiological, and radiological data.

Keywords: snake-eye; owl sign; Hirayama disease; degenerative cervical myelopathy (DCM)

---

1. Introduction

The "snake-eyes" appearance (SEA) or sign, also referred to as "owl-eyes" or "fried-eggs" sign, is a unique radiological finding appearing as bilateral hyperintense symmetric, circular or ovoid foci on T2-weighted (T2W) axial magnetic resonance imaging (MRI) sequences in the anterior horn cells of the spinal cord. It was first reported by Jenkins and Al-Mefty in 1986 [1]. The prognostic significance of this radiological finding has been debated in several articles with conflicting results [2–5].

SEA appearance is described in association with several clinical conditions like anterior spinal artery ischemia [6], chronic compressive myelopathy [7], degenerative cervical myelopathy (DCM) [4], Hirayama disease [8,9] or monomelic amyotrophy of the upper limb, amyotrophic lateral sclerosis [10], and spinal muscular atrophy [11]. The relationship between these ailments and the pathophysiology of SEA is not totally clear at present. It has been speculated that SEA is a reversible condition [3]. This claim is in contrast with its histopathology: in fact, SEA is the result of cystic necrosis at the junction of the central grey matter near the ventrolateral posterior column [12].

There is, therefore, a need to better understand SEA prognostic significance, and especially its influence on surgical outcome [13]. In this study, we evaluated the prognostic role of SEA through a systematic literature review and an analysis of our most recent patients who underwent surgical treatment.

2. Experimental Section

2.1. Literature Review

The systematic review of the literature was performed in March 2020 according to PRISMA guidelines [14]. Synthesis Without Meta-analysis (SWiM) guidelines were applied [15].

PubMed, Ovid MEDLINE, and Ovid EMBASE databases were searched using the keywords: "snake-eye myelopathy", "owl-eye myelopathy", "fried-eggs sign", "snake-eye appearance", "owl-eye appearance", and their variations. English studies published between December 1989 and December 2019 were included.

Inclusion criteria were: (1) studies with description of MRI-evident SEA myelopathy and surgery; (2) studies concerning a specific pathology related to SEA myelopathy, with patients undergoing surgical treatment; (3) studies with a clinical follow-up of surgically treated SEA patients. Exclusion criteria were: (1) absence of prognostic results about "snake-eyes sign"; (2) other radiological findings than SEA myelopathy; (3) absence of long-term follow-up.

For each study, we extracted the following baseline information: type of study, number of cases, clinical background, and prognostic value. The primary endpoint of the review was clinical outcome following surgical treatment in patients with SEA myelopathy.

2.2. Case Series

We reviewed data of all our patients with a diagnosis of DCM, who underwent surgical treatment over the past year. Patients were included in our study if (1) they had a record of "snake-eye" sign on their T2W MRI sequences and they (2) gave consent to use of their information for research purposes.

Recorded information included: baseline demographic and clinical data (age at presentation and gender; symptoms and signs at presentation), treatment strategy, outcome at discharge and follow-up. Modified Japanese Orthopedic Association Score (mJOA score) [16] and Medical Research Council (MRC) Muscle Scale [17] were adopted for pre- and post-operative neurological evaluation.

3. Results

3.1. Literature Review

A total of 77 papers were identified after duplicates removal. After title and abstract analysis, 40 articles were identified for full-text analysis. Eligibility was ascertained for three articles. PRISMA flow chart is shown in Scheme 1. SWiM scheme is reported in Table 1.

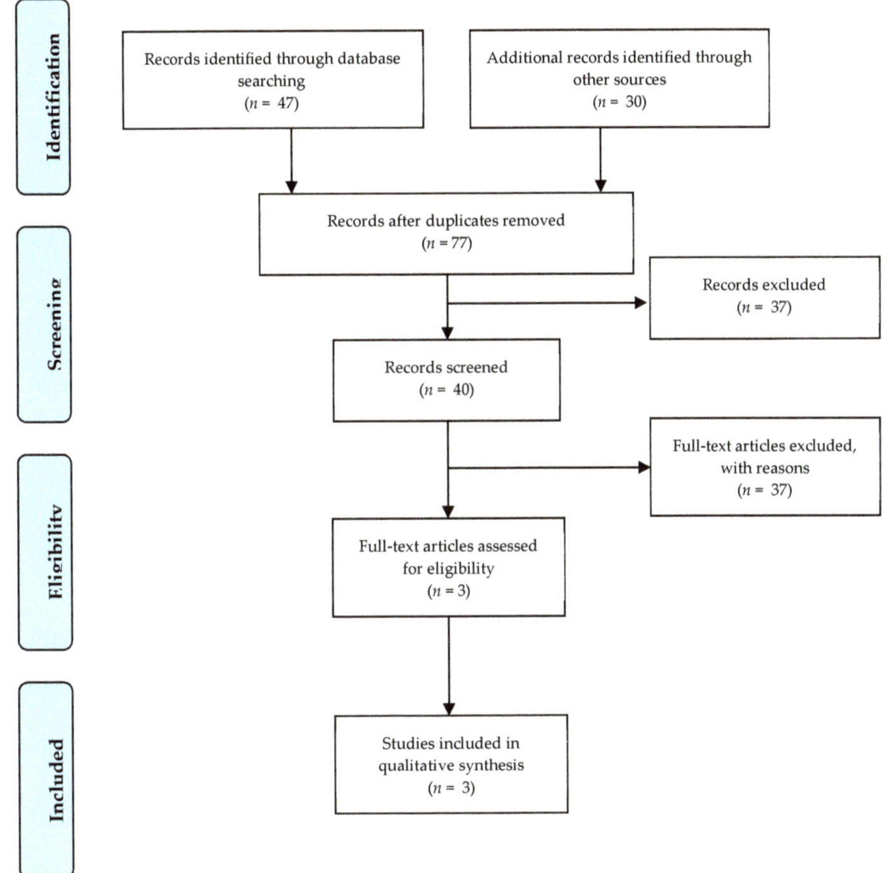

Scheme 1. Literature review process, according to PRISMA guidelines.

Choi (2005) [4] reported 47 retrospective cases of DCM, Mizuno (2003) [12] described a total of 144 retrospective cases of degenerative cervical myelopathy with a mean pre-operative mJOA score of 10.8, and Zhang (2010) [2] reported 106 retrospective cases with a diagnosis of DCM with a mean pre-operative mJOA of 8.70.

In detail, subgroup analysis reported a total of 81 patients with ossification of posterior longitudinal ligament (OPLL).

Regarding DCM, the "snake-eye" appearance was regarded as a negative prognostic factor in 144 cases (48.5%). In particular, in Mizuno's study, the improvement ratio determined by the JOA score was 32.2% in SEA (mean post-operative mJOA score of 12.9), 47.1% in NSEA, and 50% ($p < 0.01$) in control cases in which high signal intensity was absent.

Table 1. Synthesis Without Meta-analysis (SWiM) in systematic reviews.

Methods	Item Description
Grouping studies for synthesis	(1a) There is a need to better understand the influence on surgical outcome of "snake-eyes" appearance (SEA) [13]. We grouped patients undergoing surgery of the cervical spine with description of magnetic resonance imaging (MRI)-evident SEA myelopathy. (1b) We extracted data about neurological outcome following surgical procedure and pathophysiological details that could aid in understanding the natural course of SEA.
Describe the standardized metric and transformation method used	Three studies reported data suitable for descriptive statistics. Only one study reported a precise p-value. Studies were classified based on whether they showed a reduction in the outcome measure, no effect or an increase in the outcome measure following antibiotic treatment.
Describe the synthesis method	No formal statistics were adopted by the lack of sufficient data about surgical outcome. Only descriptive statistics regarding the post-operative neurological status in SEA patients were reported.
Criteria used to prioritize results for summary and synthesis	Our inclusion criteria were: (1) studies with description of MRI-evident SEA myelopathy and surgery; (2) studies concerning a specific pathology related to SEA myelopathy, with patients undergoing surgical treatment; (3) studies with a clinical follow-up of surgically treated SEA patients.
Investigation of heterogeneity in reported effects	We explored heterogeneity visually using tables, by comparing the effect sizes of studies grouped according to potential effect modifiers. These included: baseline neurological status (i.e., Modified Japanese Orthopedic Association Score (mJOA score)); pathological condition linked to SEA (e.g., cervical spondylotic myelopathy, degenerative cervical myelopathy (DCM), etc.); study design (retrospective studies).
Certainty of evidence	Two review authors (L.Z. and G.S.) independently assessed the certainty of evidence (high, moderate, low, and very low) using the five GRADE considerations (risk of bias, consistency of effect, imprecision, indirectness, and publication bias) for each of the following outcomes to draw conclusions about certainty of the evidence. We resolved disagreements on certainty ratings by discussion and provided justification for decisions to downgrade or upgrade ratings using table footnotes.
Data presentation method	We reported the synthesis of results of the included studies classified by number of patients, type of the study, condition (DCM, cervical spondylotic myelopathy) and pre- and post-operative mJOA score evaluating interventions against 'no-intervention' control groups.
Results	
Reporting results	Regarding DCM, the "snake-eye" appearance was regarded as a negative prognostic factor in 144 cases. In particular, in Mizuno's study, the improvement ratio determined by JOA score was 32.2% in SEA (mean post-operative mJOA score of 12.9), 47.1% in non snake-eye appearance (NSEA), and 50% ($p < 0.01$) in control cases in which high signal intensity was absent.
Limitation of the synthesis	The main limitations were the lack of randomized controlled trials (RCTs). Moreover, a thorough literature review shows an inconsistency of results about the prognostic significance of SEA in surgical and non-surgical patients and the pathogenetic mechanism is not completely understood and research is still ongoing.

3.2. Case Series

Case 1. A 21-year-old man presented with a one-year history of numbness in the upper limbs with severe loss of hand sensation, especially on the right side. The neurological examination documented upper extremities hypoesthesia and pain, especially in the right arm, with no clear root distribution (mJOA upper extremity sensory subscore 1 of 3) and mild weakness of wrist extensor muscles (MRC grade 4/5), again worse on the right side, with sporadic dropping of objects (mJOA upper extremity motor subscore 4 of 5). No other neurological deficits were detected with a total mJOA score of 15/18. His past medical history included a traumatic injury secondary to hyperflexion of the cervical spine, causing transient acute tetraplegia and distal sensory loss, when he was 4 years old. Cervical spine MRI documented SEA at C5–C6 without stenosis of the vertebral canal. Dynamic flexion MRI showed reduction in the spinal canal diameter with subsequent medullary compression, especially at the C5–C6 level (Figure 1).

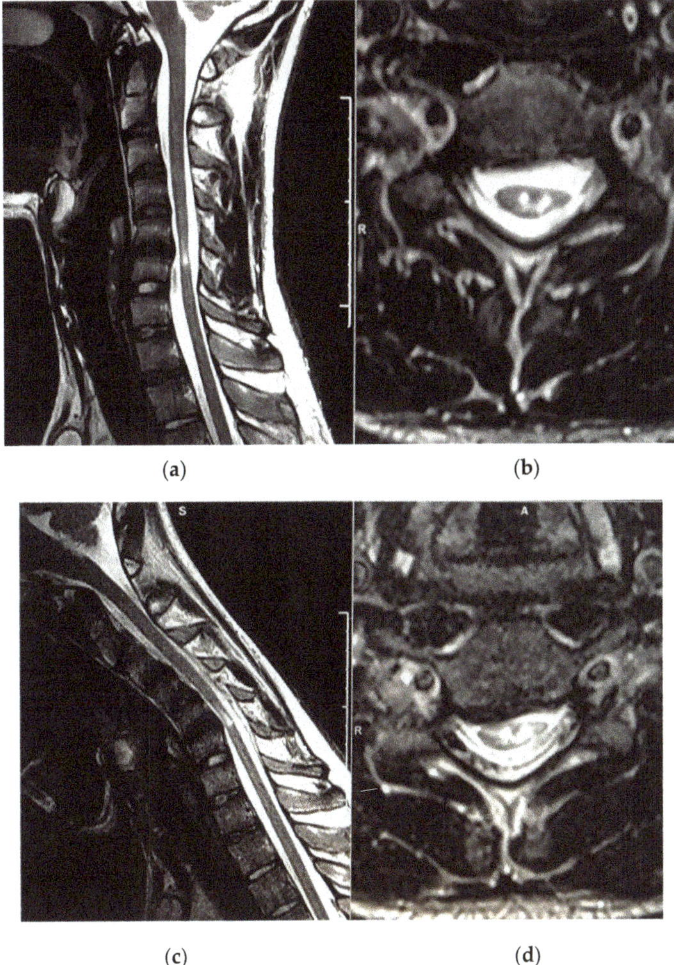

Figure 1. Dynamic sagittal and axial cervical T2-MRI scan: (**a**) sagittal T2-MRI scan with visible myelopathy at C5–C6 level; (**b**) axial T2-MRI scan with an already visible SEA appearance; (**c**) sagittal T2-MRI scan in flexion, showing reduced spinal cord canal diameter, with subsequent spinal cord compression; (**d**) axial T2-MRI scan in flexion showing SEA at C5–C6 level and spinal cord compression.

An electromyography (EMG) of the upper limbs showed signs of bilateral chronic motor axonal neuropathy in C6 myotome; low amplitude pre-operative motor evoked potentials (MEPs) were detected. The patient underwent an anterior discectomy and fusion at the C5–C6 level with an intersomatic cage. Intra-operative MEP monitoring was performed (Figure 2).

The post-operative clinical course was uneventful with no evidence of new neurological deficits. At the post-operative neurological examination, the patient had a mJOA score of 17 over 18, with an upper extremity motor improvement of 1 point and an upper extremity sensory improvement of 1 point. Wrist extensor muscles weakness resolved completely (MRC grade 5/5).

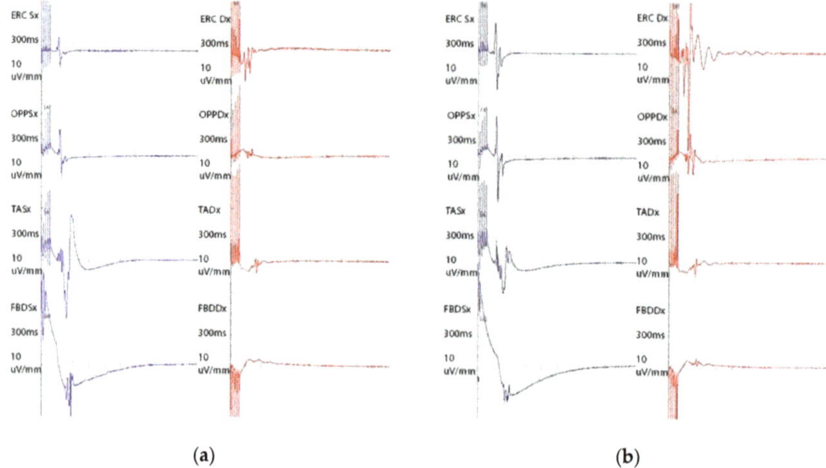

(a)   (b)

Figure 2. Intra-operative motor evoked potentials (MEPs) prior (**a**) and following (**b**) stabilization. electromyography (EMG) needles record from a hemisphere of about 1 mm. Within this volume there are some 100 muscle fibers. EMG amplitude is the sum of the electric potential differences within a muscle relating to all the active motor units in the vicinity of the electrodes on the skin [18]. Trans-cranial stimulation was performed to elicit MEPs recorded with sub-dermal needle electrodes placed bilaterally in the biceps brachii (BB), extensor radialis carpi (ERC), opponens pollicis (OPP), tibialis anterior (TA), and flexor digitorum brevis (FBD). Prior to stabilization, MEPs were elicited from all monitored muscles on the right side (red) and only from BB, ERC, and TA on the left side (blue) because of technical reasons. Following stabilization, MEPs were essentially unchanged.

Case 2. A 44-year-old man suffered a severe traumatic brain injury that required decompressive craniectomy and subsequent cranioplasty. Years later he developed arm cramps, and he was subjected to a cervical MRI scan showing a post-traumatic anterior pseudomeningocele extending from C2 to C5. He underwent multiple lumbar punctures for cerebrospinal fluid (CSF) drainage and even a spinal-peritoneal shunt, which temporarily improved his symptoms, as previously suggested [14]. However, after some time, the pain recurred, along with progressive diparesis (MRC grade 3/5 for both proximal and distal movements) and hypoesthesia, which severely affected his quality of life. Neurological evaluation detected a mJOA score of 13/18 (2/5 upper motor extremity subscore, 1/3 upper sensory extremity subscore). A new MRI scan showed an extension of the already known pseudomeningocele and a new-onset cervical snake-eyes myelopathy at the C5–C6 level (Figure 3).

Superficial upper extremities EMG confirmed denervation in the upper of both arms and low arm myotomes; low-amplitude pre-operative MEPs were detected. It was decided to perform a C3–C7 spine posterior decompression and stabilization. During surgery the patient's intraoperative MEP did not show any worsening compared to the preoperative ones (Figure 4).

Surgery was uneventful, and, at six months outpatient follow-up, the patient regained significant strength in his arms, especially distally with a MRC grade 4/5 and a mJOA score of 16/18, gaining two points on the upper limbs motor scale.

Case 3. A 56-year-old woman reported pain in both arms radiating to her hands, especially on the right side, for about five months. The patient's medical history was unremarkable. She denied recent or past trauma. More recently, she also reported the development of grip loss in her right hand, which affected her daily activities. Neurological examination detected weakness of distal right arm movements (MRC grade 4/5) with mJOA score of 15/18 (upper extremity motor subscore 4/5, upper extremity sensory subscore 1/3). Cervical MRI scan showed cervical spondylosis with associated snake-eyes myelopathy at C5–C6 level (Figure 5).

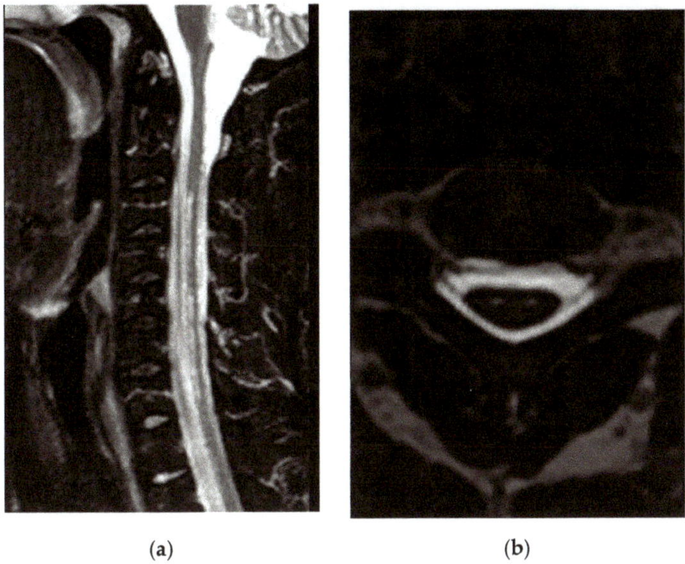

(a)   (b)

Figure 3. Pre-operative cervical T2-MRI scan: (**a**) sagittal view, showing an anterior pseudomeningocele extending from C3 to T5 with subsequent central canal stenosis and T2-high cord signal consistent with myelomalacia; (**b**) axial view, showing the edematous and T2-hyperintense anterior grey columns, with the characteristic snake-eyes sign.

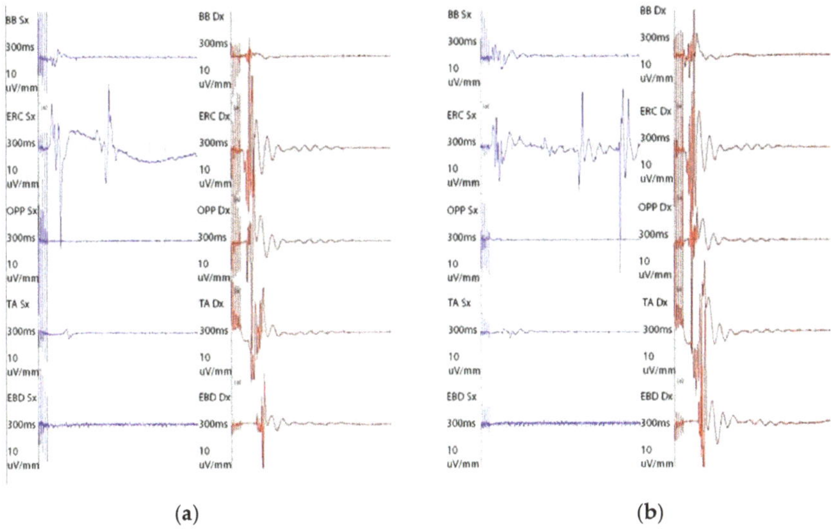

(a)   (b)

Figure 4. Intra-operative motor evoked potentials (MEPs) prior (**a**) and following (**b**) stabilization. Transcranial stimulation was performed to elicit MEPs recorded with sub-dermal needle electrodes placed bilaterally in the extensor radialis carpi (ERC), opponens pollicis (OPP), tibialis anterior (TA), and flexor digitorum brevis (FBD). Prior to stabilization lower amplitude MEPs were elicited on the right side (red) compared to the left side (blue). Following stabilization higher amplitude MEPs were elicited from the upper limbs especially on the right side.

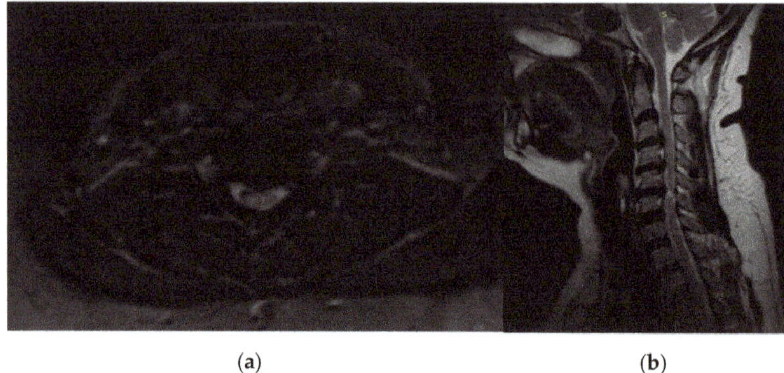

(a)   (b)

Figure 5. Pre-operative cervical T2-MRI scan: (**a**) axial view, showing cervical spondylosis with associated snake-eyes myelopathy at C5–C6 level; (**b**) sagittal view, showing the cervical canal stenosis with compression of the spinal cord.

The patient underwent a C3–C6 laminectomy and cervical arthrodesis. Surgery was uneventful. At discharge pain was reduced, especially in the right arm. The strength in her right hand was completely recovered (MRC grade 5/5 for distal arm movements) at the six month outpatient follow-up, with a mJOA score of 17/18. Post-op MRI scan is shown in Figure 6.

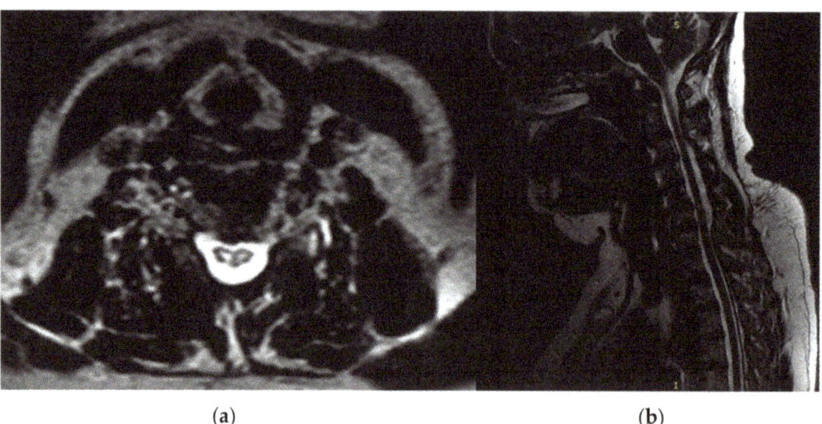

(a)   (b)

Figure 6. Post-operative cervical T2-MRI scan: (**a**) axial view through the C5–C6 level, showing resolution of the spinal canal stenosis, with persistence of the SEA signal; (**b**) sagittal view, showing resolution of the spinal cord compression.

4. Discussion

SEA was first presented in a computer tomography (CT) myelography study of seven DCM patients in 1986 [1]. Subsequent anatomopathological studies confirmed that the main modifications were cystic necrosis at the junction of the central grey matter and the posterior ventrolateral column, combined with cell loss in the anterior horn [12]. SEA was reported in other forms of myelopathy too. A clinical randomized trial [8] showed that SEA myelopathy appears during the late stage of Hirayama disease, considered as an anterior horn disorder resulting from local ischemia, triggered by arterial compression from an anterior shifting of the posterior cervical dura upon neck flexion [12].

Undoubtedly, chronic mechanical compression and vascular insufficiency can be among the main promoters of SEA [8,18]. However, the pathogenetic mechanism is not completely understood and research is still ongoing. Although there is no clear data about the exact prevalence and incidence of SEA, some studies suggest that it is much more common than it might be believed [5,8,12].

A thorough literature review shows an inconsistency of results about the prognostic significance of SEA in surgical and non-surgical patients.

According to some studies, SEA does not affect the prognosis of patients who underwent corpectomy and fusion for treatment of DCM [4,5]. Another literature review stated that intense T2W SEA is associated with poorer surgical outcome in patients with DCM, while T2W SEA post-operative regression correlates with better functional outcomes [13]. No report about T1W hypointensity is reported about SEA

Li and Remmel stated that SEA is an irreversible lesion and a predictor of poor prognosis [19]. Mizuno et al. [12] assumed that SEA is an unfavorable prognostic factor for the recovery of upper extremity motor strength and that this is related to neuronal loss in the anterior horn.

A 2015 literature review [20] suggested that SEA can help in the differential diagnosis of spinal cord ischemia, indicating anterior horns infarction caused by anterior spinal artery ischemia.

The pathogenesis of the "snake-eye" myelopathy might be a matter of some debate, but it is interesting that a cervical hypermobility could lead to an anterior compression in flexion, in the absence of a spinal canal stenosis and compression in a neutral position. The institutional Case 1 seems to show this pathogenic mechanism: a dynamic MRI study was useful for a correct therapeutic decision. The surgical strategy, in this type of patient, might be a valid choice.

Regarding the follow-up data, especially the radiological one, the entity of spinal cord damage and therefore the reversibility of the "snake eyes" sign is very difficult to verify with a post-operative MRI study, mainly for the presence of implanted materials, that could distort the signal and preclude the identification of such a modest lesion. Of our three cases, only one already underwent post-operative MRI (Case 3), showing the persistence of the snake-eyes myelopathy, despite clinical improvement. For the other two cases, only clinical follow-up is available so far. Most importantly, all three patients had a clinical benefit from surgery, despite the radiological evidence of SEA.

The MRI picture of "snake eyes" has always been described in relation to a clinical picture of myelopathy, but another relevant problem is related to those cases that have a "snake-eyes" MRI picture and are asymptomatic or pauci-symptomatic like Case 1. Low-intensity signal on T1WI is considered as a sign of advanced disease due to a significant neural tissue damage and correlates with poor post-operative neurological outcome [21]. Signal changes in T1WI usually appear with an increase in T2WI [22]. T2 hyperintensity in isolation cannot predict a worse post-operative outcome: the combination of both signal alteration on T1 and T2, and the presence of long segments with T2 hyperintensity are better correlated with negative neurological outcome after surgery [23].

For other myelopathies, it is clear that there may be signs of hyperintensity in MRI that are not related to a clinical myelopathy and in these cases it is not clear what needs to be done [24]. Unfortunately, given our small case series, the less severely affected pre-operative disability in respect of cases presented in literature and the fact that SEA is not a frequent entity presenting in a variety of heterogenous diseases, it was not possible to determine its independent prognostic value.

Nevertheless, it is widely accepted that the baseline neurological status is the strongest predictor of post-operative outcome.

In addition, older age is related to a worse outcome after surgery [16]. Other clinical factors such as body mass index and baseline severity score are not predictive of complications [25]. Duration of symptoms is not considered uniquely as a negative clinical factor associated with a negative outcome after surgery [16,25] We suggest that, from the pathophysiological point of view, a more severe cervical myelopathy and a longer duration of symptoms could associate with different histological damages in the spinal cord that could be reversible or not. Therefore, to achieve a good clinical outcome, it is mandatory to identify early clinical signs of cervical myelopathy [26]. In this regard, different outcome measures are reported in the literature regarding functional impairment (mJOA scale), disability (Nurick scale and Neck Disability Index), and generic short form health survey (SF-36 scale). Finally, new neurophysiological tests, like contact-heat evoked potentials, could be of help in early detection of cervical myelopathy; in fact, they seem to exhibit a superior sensitivity compared to somato-sensory evoked potentials in detecting spinal cord ischemia caused by compression of the anterior spinal artery. Concerning this, the application of classical neurophysiological techniques might be a limitation of this study [27].

5. Limitations of the Study

Our study suffers from several limitations. As for the analysis of the literature, few studies about DCM take into consideration surgical prognosis, and most of them have a suboptimal or no follow-up period. In addition, most studies presented an insufficient number of patients to reach the possibility of statistical inference on the general population. The variety of different pathologies that have been related to SEA myelopathy, an element that causes further difficulties in prognostic analysis, must also be considered. Furthermore, the term SEA is used unevenly in the scientific community and this may have led to loss of information. To partially compensate for this, a large number of synonyms for SEA have been used in the searching process, trying to cover the full range of terms used in the literature to describe this radiological sign. As for our case series we were unable to achieve optimal follow-up on all patients.

6. Conclusions

"Snake eyes" myelopathy represents a rare form of myelopathy with a prognosis that is generally defined as unfavorable. Its pathophysiology is still unclear, and its frequency might be greater than previously thought.

The literature review and personal experience of surgically treated cases shows that SEA represents a negative surgical prognosis sign in a minority (22–28%) of patients, but the baseline neurological status remains crucial to determine patients' outcome.

Author Contributions: L.Z., C.M.Z., E.A., G.S., S.S., L.D.M., L.Q., U.L., M.F., R.G., D.L., M.M.F., and D.L. collected, analyzed, and interpreted the patient data; R.B. and M.F. performed the surgical procedures; R.B., M.F., U.L., L.Q., M.M.F., D.L., L.Z., and C.M.Z. followed the patients during the hospital course and performed follow-up analysis. All authors have read and agree to the published version of the manuscript.

Funding: This research received no external funding.

Conflicts of Interest: The authors declare no conflict of interest.

References

1. Jinkins, J.R.; Bashir, R.; Al-Mefty, O.; Al-Kawi, M.Z.; Fox, J.L. Cystic necrosis of the spinal cord in compressive cervical myelopathy: Demonstration by Iopamidol CT-myelography. *Am. J. Roentgenol.* **1986**, *147*, 767–775. [CrossRef] [PubMed]
2. Zhang, Y.Z.; Shen, Y.; Wang, L.F.; Ding, W.Y.; Xu, J.X.; He, J. Magnetic resonance T2 image signal intensity ratio and clinical manifestation predict prognosis after surgical intervention for cervical spondylotic myelopathy. *Spine* **2010**, *35*. [CrossRef] [PubMed]
3. Takahashi, M.; Yamashita, Y.; Sakamoto, Y.; Kojima, R. Chronic cervical cord compression: Clinical significance of increased signal intensity on MR images. *Radiology* **1989**, *173*, 219–224. [CrossRef] [PubMed]
4. Choi, S.; Lee, S.H.; Lee, J.Y.; Choi, W.G.; Choi, W.C.; Choi, G.; Jung, B.; Lee, S.C. Factors affecting prognosis of patients who underwent corpectomy and fusion for treatment of cervical ossification of the posterior longitudinal ligament: Analysis of 47 patients. *J. Spinal Disord. Tech.* **2005**, *18*, 309–314. [CrossRef] [PubMed]
5. Li, H.; Jiang, L.S.; Dai, L.Y. A review of prognostic factors for surgical outcome of ossification of the posterior longitudinal ligament of cervical spine. *Eur. Spine J.* **2008**, *17*, 1277–1288. [CrossRef]
6. Bekci, T.; Yucel, S.; Aslan, K.; Gunbey, H.P.; Incesu, L. Snake eye appearance on a teenage girl with spontaneous spinal ischemia. *Spine J.* **2015**, *15*, e45. [CrossRef]

7. Tan, G.; Iwasawa, T.; Ozawa, Y.; Yoshida, T.; Matsubara, S. Snake eye sign on MR images of the cervical spine with cervical spondylosis and the clinical significance. *Yokohama Igaku* **1998**, *49*, 561–566.

8. Xu, H.; Shao, M.; Zhang, F.; Nie, C.; Wang, H.; Zhu, W.; Xia, X.; Ma, X.; Lu, F.; Jiang, J. Snake-eyes appearance on mri occurs during the late stage of hirayama disease and indicates poor prognosis. *Biomed Res. Int.* **2019**, *2019*. [CrossRef]

9. Sarawagi, R. Hirayama Disease: Imaging Profile of Three Cases Emphasizing the Role of Flexion MRI. *J. Clin. Diagn. Res.* **2014**. [CrossRef]

10. Shukla, R. Owl's eye sign: A rare neuroimaging finding in flail arm syndrome. *Neurology* **2015**, *85*, 1996.

11. Hsu, C.F.; Chen, C.Y.; Yuh, Y.S.; Chen, Y.H.; Hsu, Y.T.; Zimmerman, R.A. MR findings of Werdnig-Hoffmann disease in two infants. *Am. J. Neuroradiol.* **1998**, *19*, 550–552.

12. Mizuno, J.; Nakagawa, H.; Inoue, T.; Hashizume, Y. Clinicopathological study of "snake-eye appearance" in compressive myelopathy of the cervical spinal cord. *J. Neurosurg.* **2003**, *99*, 162–168. [CrossRef] [PubMed]

13. Vedantam, A.; Rajshekhar, V. Does the type of T2-weighted hyperintensity influence surgical outcome in patients with cervical spondylotic myelopathy? A review. *Eur. Spine J.* **2013**, *22*, 96–106. [CrossRef] [PubMed]

14. Shamseer, L.; Moher, D.; Clarke, M.; Ghersi, D.; Liberati, A.; Petticrew, M.; Shekelle, P.; Stewart, L.A.; Altman, D.G.; Booth, A.; et al. Preferred reporting items for systematic review and meta-analysis protocols (prisma-p) 2015: Elaboration and explanation. *BMJ* **2015**, *349*, 1–25. [CrossRef]

15. Campbell, M.; McKenzie, J.E.; Sowden, A.; Katikireddi, S.V.; Brennan, S.E.; Ellis, S.; Hartmann-Boyce, J.; Ryan, R.; Shepperd, S.; Thomas, J.; et al. Synthesis without meta-analysis (SWiM) in systematic reviews: Reporting guideline. *BMJ* **2020**, *368*, l6890. [CrossRef] [PubMed]

16. Kato, S.; Oshima, Y.; Oka, H.; Chikuda, H.; Takeshita, Y.; Miyoshi, K.; Kawamura, N.; Masuda, K.; Kunogi, J.; Okazaki, R.; et al. Comparison of the Japanese Orthopaedic Association (JOA) score and modified JOA (mJOA) score for the assessment of Cervical Myelopathy: A multicenter observational study. *PLoS ONE* **2015**, *10*. [CrossRef]

17. Paternostro-Sluga, T.; Grim-Stieger, M.; Posch, M.; Schuhfried, O.; Vacariu, G.; Mittermaier, C.; Bittner, C.; Fialka-Moser, V. Reliability and validity of the Medical Research Council (MRC) scale and a modified scale for testing muscle strength in patients with radial palsy. *J. Rehabil. Med.* **2008**, *40*, 665–671. [CrossRef]

18. Ulrich, A.; Min, K.; Curt, A. High sensitivity of contact-heat evoked potentials in "snake-eye" appearance myelopathy. *Clin. Neurophysiol.* **2015**, *126*, 1994–2003. [CrossRef] [PubMed]

19. Li, Y.; Remmel, K. A case of monomelic amyotrophy of the upper limb: MRI findings and the implication on its pathogenesis. *J. Clin. Neuromuscul. Dis.* **2012**, *13*, 234–239. [CrossRef]

20. Vargas, M.I.; Gariani, J.; Sztajzel, R.; Barnaure-Nachbar, I.; Delattre, B.M.; Lovblad, K.O.; Dietemann, J.L. Spinal cord ischemia: Practical imaging tips, pearls, and pitfalls. *Am. J. Neuroradiol.* **2015**, *36*, 825–830. [CrossRef]

21. Zileli, M.; Borkar, S.A.; Sinha, S.; Reinas, R.; Alves, Ó.L.; Kim, S.H.; Pawar, S.; Murali, B.; Parthiban, J. Cervical Spondylotic Myelopathy: Natural Course and the Value of Diagnostic Techniques -WFNS Spine Committee Recommendations. *Neurospine* **2019**, *16*, 386–402. [CrossRef] [PubMed]

22. Tetreault, L.A.; Dettori, J.R.; Wilson, J.R.; Singh, A.; Nouri, A.; Fehlings, M.G.; Brodt, E.D.; Jacobs, W.B. Systematic review of magnetic resonance imaging characteristics that affect treatment decision making and predict clinical outcome in patients with cervical spondylotic myelopathy. *Spine* **2013**, *38*, S89–S110. [CrossRef] [PubMed]

23. Rampersaud, Y.R.; Bidos, A.; Fanti, C.; Perruccio, A.V. The Need for Multidimensional Stratification of Chronic Low Back Pain LBP. *Spine* **2017**, *42*, E1318–E1325. [CrossRef] [PubMed]

24. Fontanella, M.M.; Fazio, M.; Francione, A.; Bacigaluppi, S.; Griva, F.; Visocchi, M.; Panciani, P.P.; Bergomi, R.; Spena, G. Pre-symptomatic cervical myelopathy: Should we operate or should we observe? What is the chance of spinal cord injury from an accident? *J. Neurosurg. Sci.* **2014**, *58*, 15–22.

25. Tetreault, L.; Ibrahim, A.; Cote, P.; Singh, A.; Fehlings, M.G. A systematic review of clinical and surgical predictors of complications following surgery for degenerative cervical myelopathy. *J. Neurosurg. Spine* **2016**, *24*, 77–99. [CrossRef]

26. Jefferson, R.W.; Lindsay, A.T.; Jun, K.; Mohammed, F.S.; James, S.H.; Thomas, M.; Samuel, C.; Michael, G.F. State of the Art in Degenerative Cervical Myelopathy: An Update on Current Clinical Evidence. *Neurosurgery* **2017**, *80*, S33–S45. [CrossRef]
27. Jutzeler, C.R.; Ulrich, A.; Huber, B.; Rosner, J.; Kramer, J.L.K.; Curt, A. Improved Diagnosis of Cervical Spondylotic Myelopathy with Contact Heat Evoked Potentials. *J. Neurotrauma* **2017**, *34*, 2045–2053. [CrossRef]

Journal of
Clinical Medicine

Article

Degenerative Cervical Myelopathy: How to Identify the Best Responders to Surgery?

Rocco Severino [1], Aria Nouri [2,*] and Enrico Tessitore [2]

1   Division of Neurosurgery, IRCCS Neuromed, 86077 Pozzilli (IS), Italy; severinorocco@gmail.com
2   Department of Neurosurgery, Hôpitaux Universitaires de Genève (HUG), 1205 Geneva, Switzerland; enrico.tessitore@hcuge.ch
*   Correspondence: Aria.Nouri@hcuge.ch

Received: 1 February 2020; Accepted: 5 March 2020; Published: 11 March 2020

Abstract: Surgery is the only definitive treatment for degenerative cervical myelopathy (DCM), however, the degree of neurological recovery is often unpredictable. Here, we assess the utility of a multidimensional diagnostic approach, consisting of clinical, neurophysiological, and radiological parameters, to identify patients likely to benefit most from surgery. Thirty-six consecutive patients were prospectively analyzed using the modified Japanese Orthopedic Association (mJOA) score, MEPs/SSEPs and advance and conventional MRI parameters, at baseline, and 3- and 12-month postoperatively. Patients were subdivided into "normal" and "best" responders (<50%, ≥50% improvement in mJOA), and correlation between Diffusion Tensor Imaging (DTI) parameters, mJOA, and MEP/SSEP latencies were examined. Twenty patients were "best" responders and 16 were "normal responders", but there were no statistical differences in age, T2 hyperintensity, and midsagittal diameter between them. There was a significant inverse correlation between the MEPs central conduction time and mJOA in the preoperative period ($p = 0.0004$), and a positive correlation between fractional anisotropy (FA) and mJOA during all the phases of the study, and statistically significant at 1-year (r = 0.66, $p = 0.0005$). FA was significantly higher amongst "best responders" compared to "normal responders" preoperatively and at 1-year ($p = 0.02$ and $p = 0.009$). A preoperative FA > 0.55 was predictor of a better postoperative outcome. Overall, these results support the concept of a multidisciplinary approach in the assessment and management of DCM.

Keywords: degenerative cervical myelopathy (DCM); surgical outcome; MRI; DTI; FA; ADC; signal changes spinal canal; neurophysiology; SSEP; MEP

1. Introduction

Degenerative cervical myelopathy (DCM) is typically a chronic condition, commonly involving patients older than 55 years [1], and represents the most common cause of spinal cord injury in the industrialized world [2]. The progressive reduction of spinal canal diameter due to degeneration of the cervical spine, including the vertebrae, posterior longitudinal ligament, ligamentum flavum, intervertebral disk [3], results in a compression of the spinal cord, arterial perfusion to the nervous tissue, and consequent spinal cord ischemia [4,5]. DCM can be a highly disabling condition causing motor and sensory dysfunction that ultimately result in a reduced quality of life.

The diagnosis of DCM is based on clinical examination, and subsequently confirmed using imaging, and sometimes neurophysiological techniques such as sensory (SSEPs) and motor evoked potentials (MEPs). Studies using conventional MRI have shown that specific characteristics can correlate with neurological status and surgical outcome. The most commonly studied parameters include T1-weighted hypointensity or T2-weighted hyperintensity signals of the spinal cord and the number of compressed levels. It is believed that signal changes represent a wide-ranging set of pathological sequelae. Edema and gliosis are thought to result in demyelination and Wallerian degeneration, and are typically associated with T2 hyperintensity signal changes in the absence of T1 hypointensity [6]. After prolonged compression or significant dynamic injury, myelomalacia and loss of grey matter occurs [7–10], typically reflected by T1 hypointensity signal changes. However, T2 hyperintensity presents in 58%–85% of DCM patients, but it is present in 2.3% of people in the general population as well, making it a sensitive measure for diagnosis, but limiting it in terms of predicting surgical outcome. Contrarily, T1 hypointensity has been found to be a good predictor of suboptimal surgical outcome but its low prevalence in DCM of about 20% of patients limits its clinical utility [11].

In recent years, a newer MRI technique, the diffusion tensor imaging (DTI), has demonstrated an ability to identify the degenerative changes of the compressed spinal cord even in the early phases of DCM. [12,13] In the neural tissue, DTI imaging estimates the directionality and the diffusivity of water molecules through tissues and nervous fibers through two values: The fractional anisotropy (FA) and the apparent diffusion coefficient (ADC). These scalar parameters are inversely proportional and, in a damaged and demyelinated spinal cord water molecules diffuse in all different directions resulting in lower FA and higher ADC studies values [10,14], while in the normal population FA has higher scores because of intact myelin sheaths. [14,15] Moreover, several studies have shown a correlation between preoperative FA, ADC values, and clinical condition of DCM patients [13,16].

Neurophysiological studies are also useful diagnostic tools in the detection of functional alterations in nerve conduction in DCM. Several studies have demonstrated significant alterations in both SSEPs and MEPs in DCM patients [17–19]; specifically, a predictive value for the postsurgical outcome has been shown for median nerve SSEPs [19–21].

In a previous study, we were able to demonstrate the importance of combining clinical, radiological, and neurophysiological data in the assessment of DCM patients, in order to better identify those with optimal surgical outcomes [22]. The aim of the present study is to evaluate the correlation between DTI, neurophysiological parameters and neurological status, in both the preoperative and postoperative periods, in order to define a new multidisciplinary diagnostic approach that could identify the best candidates for decompressive surgery.

2. Methods

We performed a prospective analysis of clinical, radiological, and neurophysiological data of thirty-six consecutive patients (13 males, 23 females; mean age: 57.05 years) suffering from DCM and operated between June 2012 and June 2018. Inclusion criteria are detailed in Table 1. All subjects gave their written informed consent for inclusion before they participated in the study. Research ethics board approval was obtained by the Institutional Ethical Board (NAC n. 11-194). A complete clinical, radiological, and neurophysiological evaluation of DCM, consisting of cervical MRI with DTI sequences, the modified Japanese Orthopedic Association (mJOA) score by Keller et al. [23], and neurophysiological assessments (both MEPs and SSEPs), was performed for all patients in the preoperative period and repeated at 3- and 12-months after surgery. The choice of the surgical approach depended on the predominant side of compression and on the surgeon's preferences. The posterior approach was chosen in patients with a multilevel compression and with preservation of the cervical lordosis on plain X-rays. Fusion was added in cases with radiological instability according to White and Panjabi criteria [24].

The mJOA score was used to measure the severity of DCM and segregated into patients with normal function (mJOA = 16–17), grade 1 (mJOA = 12–15), and grade 2 (mJOA = 8–11) myelopathy. We calculated the improvement of DCM using the mJOA recovery rate proposed by Hirabayashi, as follows: [(Postoperative mJOA-preoperative mJOA)/(17-preoperative mJOA) × 100] [25]. "Best responders" were identified by improvement of 50% or more in the postoperative period [26]. Patient improving below 50% or remaining stable (at least no deterioration) were defined as "normal responders".

Table 1. Inclusion and exclusion criteria.

Inclusion Criteria	Exclusion Criteria
Age: 30–81 years	Contraindication to perform MR
Cervical stenosis at MR	Epileptic patients
Need for decompressive surgery	Pregnant
Clinical signs of myelopathy	Cancer or infection
Complete follow up at three months and one year	Previous cervical spine surgery

2.1. Radiological Assessment

All patients underwent a cervical 3 Tesla MR with the following sequences: Sagittal fast spin echo T2 (FSET2) TE 102 ms, TR 3500 ms, slice thickness 3 mm, axial gradient echo T2 TE 14 ms, TR 676 ms, slice thickness 3 mm, sagittal FSE T1, TE 10 ms, TR 900 ms, slice thickness 3 mm, sagittal diffusion tensor imaging (DTI) TE 58 ms, TR 4000 ms, *b*-value 700, slice thickness 2 mm, 25 directions with calculation of the apparent diffusion coefficient (ADC) and fractional anisotropy (FA).

The images were obtained in the preoperative period and at 3- and 12-months after surgery. The FA and the ADC were measured with isometric ROIs at three levels: The surgical level, or the narrowest point of the cervical stenosis in case of multilevel compression (level 2), and the nearest noncompressed intervertebral levels above and below (level 1 and 3, respectively—Figure 1). The presence and the extension of T2 hyperintensity, and the presence of T1 hypointensity was also investigated.

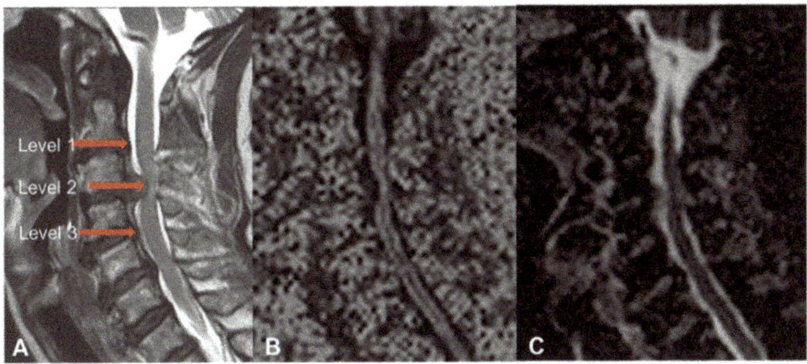

Figure 1. A: Measurement levels for fractional anisotropy (FA) and the apparent diffusion coefficient (ADC): The surgical level, or the narrowest point of the cervical stenosis in patients with multilevel compression (level 2), and the intervertebral levels above and below it (level 1 and 3). **B, C**: Diffusion tensor imaging (DTI) sequences for FA and ADC measurement, respectively.

The midsagittal diameter of the spinal canal at the site of greatest compression, was calculated on sagittal T2 images in both preoperative and postoperative controls (at three months) and used this to examine the expansion rate of the spinal canal after the surgery, using the formula: [(Postoperative–Preoperative AP diameter)/Preoperative AP diameter] × 100.

2.2. Electrophysiological Assessment

All patients had undergone electrophysiological evaluation with SSEPs and MEPs. The Quadriceps Combined Test [27] was used for the analysis of the MEPs. Furthermore, calculation of the preoperative and postoperative (3-months and 1-year) central conduction time (TCC), corrected for the age and size, and the amplitude ratio of the MEP was obtained in all patients.

For SSEPs, the peak latencies of responses were recorded at Erb's point (N9), the C2 spinous process (N13), and the scalp (N20) for the median nerve. For the tibial nerve, we calculated the latencies N8–N22 (popliteal fossa to L1) and N22 (L1 spinous process). Unfortunately, several patients did not perform a complete postoperative SSEP analysis according to our criteria as a result of a mismatch between different neurophysiological diagnostic units. Therefore, only the SSEP analysis on the preoperative data was conducted.

2.3. Statistical Analysis

Data were analyzed using an unpaired t-test to compare the mJOA, midsagittal diameter of the spinal canal, DTI parameters, and MEP/SSEP values, during all the phases of the study, in the "best responders" and "normal responders" to surgery. A Pearson correlation analysis was then performed to assess the correlation between pre and postoperative FA and ADC values, mJOA score, and MEP/SSEP values. Fischer exact tests were performed to determine a specific preoperative FA threshold that could be predictive of a better postoperative outcome. Fisher exact tests were also used to investigate if general "risk factors" (T2 hyperintensity, Tobacco use, diabetes, and clinical history >6 months) were related to worse outcome. A p-value < 0.05 was considered to be statistically significant.

3. Results

3.1. Clinical Demographics and Outcome

Twenty-six patients (72.2%) were operated by an anterior approach, including an anterior discectomy and fusion (ACDF) in 20 patients and an anterior corpectomy and fusion in six patients. Ten patients (27.3%) were treated with posterior decompression by laminectomy with or without fusion. Based on the mJOA score, there were 5/36 patients with normal function, 25/36 patients with grade 1 myelopathy, and 6/36 patients with grade 2 myelopathy. There were no patients with a preoperative grade 3 myelopathy.

The preoperative mJOA average value was 13.5, while the postoperative mean values were 14.9 at 3-months and 15.1 at 1-year, respectively. According to the Hirabayashi recovery ratio, 20 patients (55.5%) were considered "best responders"; the difference between the mJOA improvement of the "best responders" and "normal responders" patients at 1-year was statistically significant ($p = 0.001$, Figure 2). The difference between the mean age of the "best responders" and "normal responders" was not statistically significant (58.9 ± 13.2 vs. 54.6 ± 13.1, $p = 0.34$).

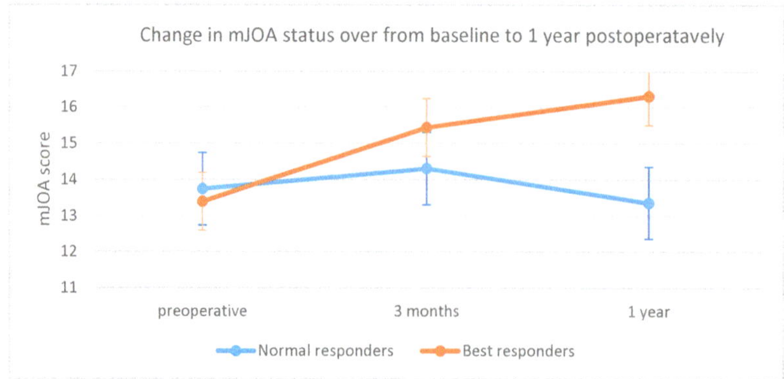

Figure 2. Difference of trends of the average modified Japanese Orthopedic Association (mJOA) scores in the "best responders" and "normal responders" patients. The improvement from the preoperative score to the 1-year value in the "best responders" group was significant ($p = 0.001$), as the difference between the 1-year values of the "best responders" (mean = 16.3) and the "normal responders" (mean = 13.3) patients ($p = 0.001$).

There was no statistical difference between the two groups concerning both age (58.9 ± 13.2 vs. 54.6 ± 13.1, p: 0.34) and the investigated "risk factors" (T2 hyperintensity, p: 0.13; smoke, p: 0.22; diabetes, p: 0.83; clinical history > 6 months, p: 0.14. Table 2).

Table 2. Differences in characteristics between best responders and normal responders.

Characteristics		Best Responders	Normal Responders	p-Value
		$n = 20$	$n=16$	
Age		58.9 ± 13.2	54.6 ± 13.1	0.34
Smoke		0.30	0.50	0.22
T2 hyperintensity		0.50	0.75	0.13
Diabetes		0.05	14.2%	0.83
Symptoms > 6 month		81.8%	0.75	0.14
Midsagittal diameter (mm)	Preoperative	5.10 ± 1.4	5.15±1.4	0.44
	Postoperative	8.98 ± 2.3	8.84±1.6	0.48
Cord expansion rate		100.1%	93.9%.	0.8
Average FA	Preoperative	0.63 ± 0.06	0.57±0.08	0.03 *
	Three months	0.62 ± 0.08	0.58±0.09	0.3
	One year	0.68 ± 0.07	0.55±0.11	0.004 *
Surgical level FA	Preoperative	0.63 ± 0.15	0.55±0.11	0.02 *
	Three months	0.62 ± 0.09	0.56±0.14	0.33
	One year	0.67 ± 0.08	0.54±0.09	0.009 *

*: Statistically signifant result.

3.2. Radiological Results

The mean midsagittal canal diameter was 5.12 ± 1.4 and 8.92 ± 2 mm in the preoperative and postoperative period, respectively, with a mean expansion rate of 97.4%. Considering our cases as "best responder" and "normal responder" patients, we found no significant differences between the average values of both postoperative midsagittal diameters (8.98 ± 2.3 vs. 8.84 ± 1.6 mm, $p = 0.8$) and expansion rates (100.1% vs. 93.9%, $p = 0.8$).

Concerning the preoperative DTI parameters, the preoperative FA values were significantly higher in the "best responders" than the "normal responders" (0.63 ± 0.06 vs. 0.57 ± 0.08, $p = 0.03$). Six patients were excluded from the postoperative analysis because they presented with artefacts on their MRIs related to implanted metallic devices. In the remaining 30 patients, the average FA value remained higher in the "best responders" group at both 3-months (0.62 ± 0.08 vs. 0.58 ± 0.09) and statistically significantly different at 1-year (0.68 ± 0.07 vs. 0.55 ± 0.11, $p = 0.004$, Table 2). Furthermore, FA at the most stenotic level was significantly lower in the "normal responder" group preoperatively and at 1-year ($p = 0.02$ and $p = 0.009$, respectively—Figure 3, Table 2).

The "best responders" group had a preoperative FA > 0.55 in 71.5% of patients compared with only 28.5% of the "normal responders". This difference was statistically significant (p = 0.014), and suggests that a preoperative FA > 0.55 can be considered as a predictor of a better postoperative outcome.

T2 hyperintensity in the preoperative MRI was found in 12/16 (75%) of the "normal responders" patients and in 10/20 (50%) in the "best responders" group; however, this difference was not significantly different ($p = 0.13$). The preoperative average FA was similar in patients with and without T2 hyperintensity (0.595 vs. 0.596, $p = 0.9$). No T1 hypointensity was evident in any of our patients.

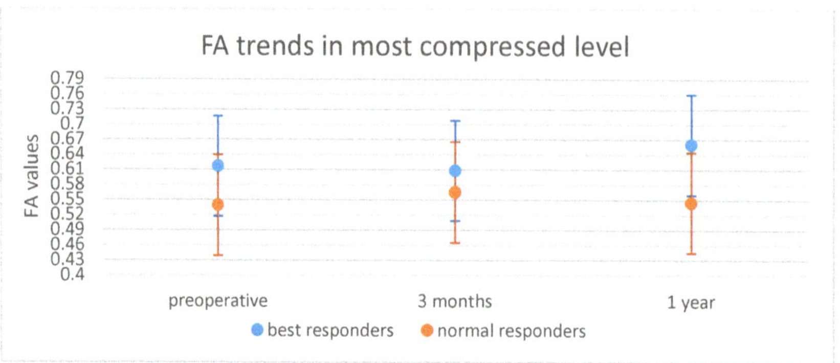

Figure 3. Differences between fractional anisotropy average values of the most compressed level in best responders (blue) and normal responders (red) patients.

Concerning the ADC, the average value between the "normal responder" group compared to the "best responders" group was higher in the preoperative period (1.54 vs. 1.40) and at the 3-month follow-up (1.40 vs. 1.27), but lower at the 1-year control (1.37 vs. 1.40 in the "best responders" group). These results did not show a statistically significant difference, and we did not consider a lower ADC as a predictive factor for good recovery.

3.3. Neurophysiological Results

Concerning the MEPs, a significant inverse correlation between the CCT and mJOA values in the preoperative period ($p = 0.0004$, R = −0.59, Figure 4) was found.

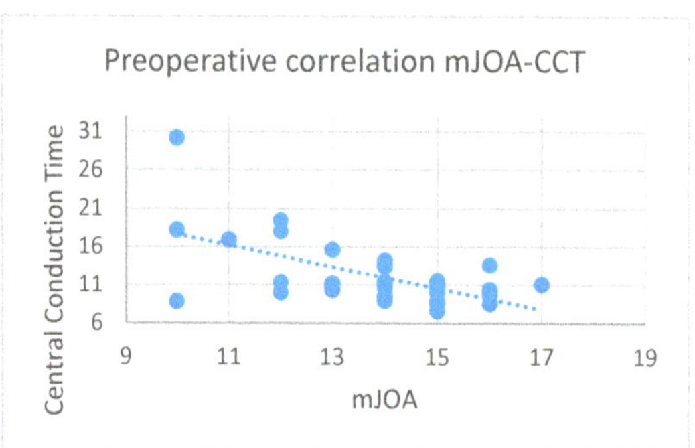

Figure 4. Preoperative abnormal values of motor evoked potentials (MEPs) were related to worse mJOA scores: This inverse correlation was statistically significant (r = −0.59, $p = 0.0004$).

We found no correlation between abnormal preoperative SSEPs and preoperative mJOA scores. Regarding the relationship between SSEP and FA values, we observed a significant inverse correlation between preoperative FA and N22, N8–N22 latencies ($p = 0.001$ and $p = 0.007$, respectively (Figure 5), there was no statistically significant correlation with the other SSEPs otherwise. The same correlation was found postoperatively but due to the mismatch of the test conducted in different diagnostic centers we were not able to conduct any statistical analysis.

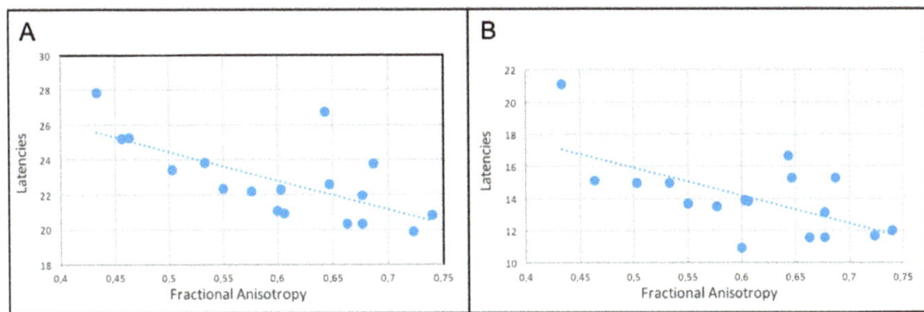

Figure 5. **A:** Inverse correlation between preoperative FA values and L1 spinous process (N22) ($p = 0.001$). **B:** Inverse correlation between preoperative FA values and popliteal fossa to L1 (N8–N22) ($p = 0.007$).

3.4. Correlation between FA, mJOA Values, and Neurophysiological Parameters

A positive correlation between FA values and corresponding mJOA scores during all the phases of the study was found. However, this correlation was significant only for the 1-year postoperative values ($p = 0.0005$, R = 0.66, Figure 6).

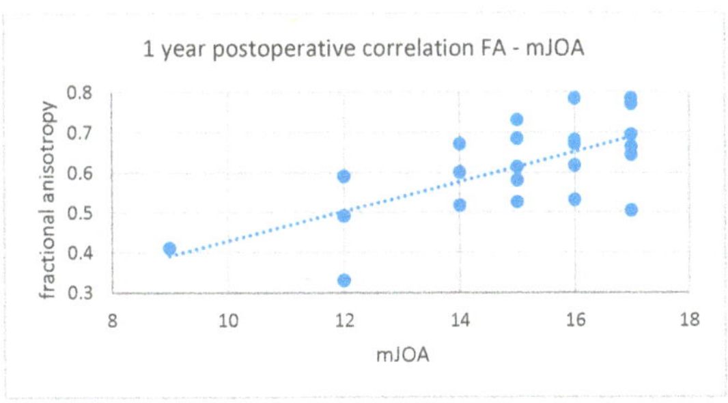

Figure 6. Positive correlation between preoperative FA and mJOA at 1 year ($p = 0.004$, r = 0.66).

Moreover, a direct correlation was found between higher preoperative FA values, and the postoperative variation of the mJOA at 1 year. This result was significant considering both the FA at the most compressed level and the average value of all the three considered levels ($p = 0.002$, r = 0.66 and $p = 0.0002$, r = 0.75, respectively—Figure 7).

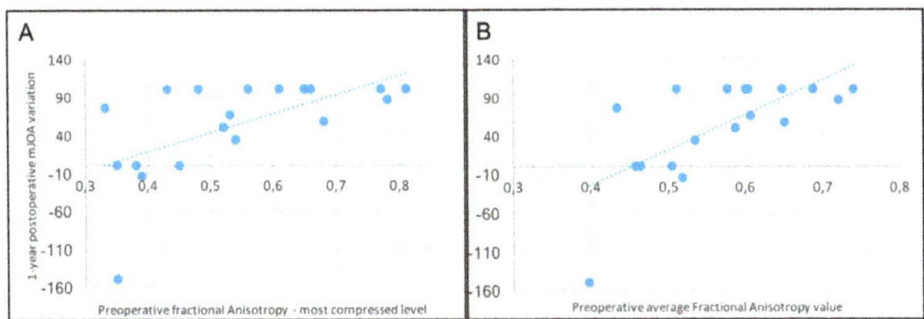

Figure 7. A: Significant correlation between preoperative FA value at the most compressed level and the 1-year postoperative variation of the mJOA ($p = 0.002$). **B:** Significant correlation between preoperative FA average value and the 1-year postoperative variation of the mJOA ($p = 0.0002$).

Concerning the relationship between FA and MEP values, an inverse correlation was found between TCC values and FA scores in the preoperative period, but this result was not statistically significant (r = −0.33; *p* = 0.08).

4. Discussion

DCM is a complex and potentially disabling condition. The time for surgical intervention is usually dictated by the degree of neurologically severity, with moderate to severely impaired patients recommended for surgery, whereas mildly impaired patients may be offered surgery or careful observation. [28] Although most patients with DCM will improve after surgery, it remains challenging to accurately predict good responders. These difficulties are due, in part, to the different pathophysiological causes of chronic cervical myelopathy such as ischemic degeneration of the neural tissue due to hypoperfusion, loss of motoneurons in the anterior horns [29], membrane damages, and conduction decline [30]. Presently, we are faced with a true *"paradigm shift"*, passing from an era where the goal of surgery for DCM was to stop the disease progression, to an era where surgery seems to be able to improve patients' status in most of the cases, having thus a favorable impact on DCM patients' quality of life [31].

Unfortunately, there is a lack of a standard diagnostic protocol that can provide prognostic information for DCM patients. Past notions about the usefulness of MRI findings in DCM, such as the presence of T2 hyperintensity, have shown to be largely nonspecific with regards to the severity of myelopathy in specific patients and the capacity of postoperative neurological improvement [8,13,32–34]. Having said this, T2 hyperintensity seems to have some potential utility when measured in terms of sagittal extent, or when skip lesions are observed [11].

It remains unclear if age impacts outcome, while some have demonstrated that age is a predictive variable, others have found no such relationship for the postsurgical recovery rate [35,36]. In our study we found that T2 hyperintensity, Tobbaco use, diabetes, and a clinical history >6 months, are not related to a worse outcome. Furthermore, multilevel compression was also not predictive of a poorer neurological outcome. While, these results are consistent with some studies found in literature [8,37–40], other studies, notably those derived from the AOSpine multicenter studies on DCM, have shown contradictory results [11,41]. Our results may partially be due to our relatively small cohort and may have been underpowered to detect statistically significant differences.

In contrast with the results reported by various authors [41–44], we found that surgical outcome was independent from both the age and the spinal canal diameter before surgery. In fact, we observed no statistical difference between the spinal canal midsagittal diameter of "best responders" and "normal responders", both in the preoperative and postoperative periods. This may be partially due to dynamic injury mechanisms that contribute to DCM and that are not necessarily influenced by canal diameter.

New MRI techniques, such as the DTI, in combination with neurophysiological assessment can help identify those patients with higher probabilities of improving after surgical decompression [22]. Similarly to previous research [45], FA in "best responders" were significantly higher than those of the "normal responders", both preoperatively and 1-year follow-up. Moreover, there was also a statistical difference in the FA values of the most stenotic level in all the phases of the study between the two groups, and a significant relationship between a preoperative FA > 0.55 at the most compressed level and a better clinical outcome (RR > 50%) at 1-year. This is consistent with the concept that FA can be used in the assessment of the degree of severity of DCM with greater accuracy. [12–14,34,43] Our results support what has been previously suggested—higher values of preoperative FA can be considered as a positive prognostic factor of functional recovery [13,34,37]. It was interesting to note that, FA values were not different between best and normal responders at 3-months, and this potentially indicates a 3-month FA MRI may be too early to detect changes.

Neurophysiological parameters have been reported as useful diagnostic tools for DCM. It has been previously shown that MEPs have a good diagnostic value in the setting of DCM, perhaps even more so than SSEPs [19]. Our study shows support to this concept in that preoperative abnormal CCT was related to a worse clinical condition and a lower mJOA ($p = 0.0004$). The significant correlation between preoperative FA and N22, N8–N22 latencies in the preoperative period are challenging to interpret—further research needs to be done in this area to better understand this relationship and how it may be useful in diagnosis and outcome prediction, or if this was a spurious finding.

Several authors stated that median SSEPs can be used for both diagnostic and prognostic purposes in DCM patients. Lyu et al. [20] stated that normal median SSEPs are related to a better prognosis; Morishita et al. [21] showed that an early improvement of the N18 was related to a good outcome at 3-months after surgery. Restuccia et al. [19] found that the SSEP improvement was related to a better clinical outcome, especially in those patients with an isolated loss of N13. These results confirm that SSEPs are a useful tool for the diagnosis of a cervical myelopathy. Unfortunately, in our study we were not able to perform a complete SSEP analysis in all our patients in the postoperative period. Nevertheless, we observed in the preoperative data, a correlation between fractional anisotropy and neurophysiological parameters and in particular tibial SSEPs, suggesting the existence of connection a between a radiological information and a clinical data.

Limitations

The greatest limitation of the present study is the relatively small cohort and the retrospective nature of our analysis. However, few studies have assessed the combination of electrophysiology in conjunction with the commonly used clinical and advanced MRI parameters. While our study did not support some of the findings of previous authors that remain supported with low evidence, our findings with regards to FA are in accordance with others [34,45,46] and strongly support the effectiveness of DTI analysis in both assessing the clinical status and predicting the surgical outcome in DCM patients.

Unfortunately, the use of different neurophysiological methods in the postoperative period did not allow us to assess the effectiveness of changes in these parameters in the postoperative setting.

Lastly, patients with preoperative mJOA scores of 17 were included; however, while these patients did not show clear neurological impairment as assessed by the mJOA, they exhibited neurophysiological evidence and/or objective clinical signs such as hyper-reflexia, Hoffman's sign.

5. Conclusions

Our results validate the concept that the current "ordinary" assessment of DCM should be upgraded with new diagnostic techniques. DTI could be considered not only a complementary diagnostic analysis, but rather a crucial tool in order to identify the best candidates to surgery. Neurophysiological parameters, in particular MEPs, correlate with the clinical condition of the patient, and could therefore be considered as an additional diagnostic tool in the preoperative period. The inclusion of DTI sequences in the preoperative study provides also prognostic information, enhancing the presurgical evaluation of DCM patients. Our findings suggest that FA values are most useful preoperatively and at 1-year follow-up, and may not be useful at 3-months postoperatively. The inclusion of electrophysiology and DTI measurements may enhance the diagnostic process and may be effective at augmenting the predictive capacity of previously described prediction models.

Author Contributions: Conceptualization and methodology: E.T. Formal analysis, data curation: R.S. Writing—original draft preparation, R.S., E.T. Writing—review and editing: E.T, A.N., R.S. Visualization, supervision: E.T., A.N. All authors have read and agreed to the published version of the manuscript.

Conflicts of Interest: The authors declare no conflict of interest.

References

1. Matz, P.G.; Anderson, P.A.; Groff, M.W.; Heary, R.F.; Holly, L.T.; Kaiser, M.G.; Mummaneni, P.V.; Ryken, T.C.; Choudhri, T.F.; Vresilovic, E.J.; et al. Cervical laminoplasty for the treatment of cervical degenerative myelopathy. *J. Neurosurg. Spine* **2009**, *11*, 157–169. [CrossRef] [PubMed]

2. Nouri, A.; Tetreault, L.; Singh, A.; Karadimas, S.K.; Fehlings, M.G. Degenerative Cervical Myelopathy: Epidemiology, Genetics, and Pathogenesis. *Spine* **2015**, *40*, E675–E693. [CrossRef] [PubMed]

3. Bernhardt, M.; Hynes, R.A.; Blume, H.W.; White, A.A. Cervical spondylotic myelopathy. *J. Bone Joint Surg. Am.* **1993**, *75*, 119–128. [CrossRef] [PubMed]

4. Gooding, M.R.; Wilson, C.B.; Hoff, J.T. Experimental cervical myelopathy. Effects of ischemia and compression of the canine cervical spinal cord. *J. Neurosurg.* **1975**, *43*, 9–17. [CrossRef] [PubMed]

5. Nurick, S. The natural history and the results of surgical treatment of the spinal cord disorder associated with cervical spondylosis. *Brain* **1972**, *95*, 101–108. [CrossRef] [PubMed]

6. Nouri, A.; Martin, A.R.; Mikulis, D.; Fehlings, M.G. Magnetic resonance imaging assessment of degenerative cervical myelopathy: A review of structural changes and measurement techniques. *Neurosurg. Focus* **2016**, *40*, E5. [CrossRef]

7. Mummaneni, P.V.; Kaiser, M.G.; Matz, P.G.; Anderson, P.A.; Groff, M.; Heary, R.; Holly, L.; Ryken, T.; Choudhri, T.; Vresilovic, E.; et al. Preoperative patient selection with magnetic resonance imaging, computed tomography, and electroencephalography: Does the test predict outcome after cervical surgery? *J. Neurosurg. Spine* **2009**, *11*, 119–129. [CrossRef]

8. Tetreault, L.; Dettori, J.R.; Wilson, J.R.; Singh, A.; Nouri, A.; Fehlings, M.G.; Brodt, E.D.; Jacobs, W.B. Systematic review of magnetic resonance imaging characteristics that affect treatment decision making and predict clinical outcome in patients with cervical spondylotic myelopathy. *Spine* **2013**, *38*, S89–S110. [CrossRef]

9. Karpova, A.; Arun, R.; Cadotte, D.W.; Davis, A.M.; Kulkarni, A.V.; O'higgins, M.; Fehlings, M.G. Assessment of spinal cord compression by magnetic resonance imaging—can it predict surgical outcomes in degenerative compressive myelopathy? A systematic review. *Spine* **2013**, *38*, 1409–1421. [CrossRef]

10. Ellingson, B.M.; Salamon, N.; Holly, L.T. Advances in MR imaging for cervical spondylotic myelopathy. *Eur. Spine J.* **2015**, *24* (Suppl. 2), 197–208. [CrossRef]

11. Nouri, A.; Martin, A.R.; Kato, S.; Reihani-Kermani, H.; Riehm, L.E.; Fehlings, M.G. The Relationship Between MRI Signal Intensity Changes, Clinical Presentation, and Surgical Outcome in Degenerative Cervical Myelopathy: Analysis of a Global Cohort. *Spine* **2017**, *42*, 1851–1858. [CrossRef] [PubMed]

12. Kerkovsky, M.; Bednarik, J.; Dusek, L.; Šprláková-Puková, A.; Urbánek, I.; Mechl, M.; Válek, V.; Kadanka, Z. Magnetic resonance diffusion tensor imaging in patients with cervical spondylotic spinal cord compression: Correlations between clinical and electrophysiological findings. *Spine* **2012**, *37*, 48–56. [CrossRef] [PubMed]

13. Jones, J.G.A.; Cen, S.Y.; Lebel, R.M.; Hsieh, P.C.; Law, M. Diffusion tensor imaging correlates with the clinical assessment of disease severity in cervical spondylotic myelopathy and predicts outcome following surgery. *AJNR Am. J. Neuroradiol.* **2013**, *34*, 471–478. [CrossRef]

14. Guan, X.; Fan, G.; Wu, X.; Gu, G.; Gu, X.; Zhang, H.; He, S. Diffusion tensor imaging studies of cervical spondylotic myelopathy: A systemic review and meta-analysis. *PLoS ONE* **2015**, *10*, e0117707. [CrossRef] [PubMed]

15. Tsuchiya, K.; Katase, S.; Fujikawa, A.; Hachiya, J.; Kanazawa, H.; Yodo, K. Diffusion-weighted MRI of the cervical spinal cord using a single-shot fast spin-echo technique: Findings in normal subjects and in myelomalacia. *Neuroradiology* **2003**, *45*, 90–94. [CrossRef] [PubMed]

16. Budzik, J.-F.; Balbi, V.; Le Thuc, V.; Duhamel, A.; Assaker, R.; Cotten, A. Diffusion tensor imaging and fibre tracking in cervical spondylotic myelopathy. *Eur. Radiol.* **2011**, *21*, 426–433. [CrossRef]

17. Brunhölzl, C.; Claus, D. Central motor conduction time to upper and lower limbs in cervical cord lesions. *Arch. Neurol.* **1994**, *51*, 245–249. [CrossRef]

18. Tavy, D.L.; Wagner, G.L.; Keunen, R.W.; Wattendorff, A.R.; Hekster, R.E.; Franssen, H. Transcranial magnetic stimulation in patients with cervical spondylotic myelopathy: Clinical and radiological correlations. *Muscle Nerve* **1994**, *17*, 235–241. [CrossRef]

19. Restuccia, D.; Di Lazzaro, V.; Lo Monaco, M.; Evoli, A.; Valeriani, M.; Tonali, P. Somatosensory evoked potentials in the diagnosis of cervical spondylotic myelopathy. *Electromyogr. Clin. Neurophysiol.* **1992**, *32*, 389–395.

20. Lyu, R.K.; Tang, L.M.; Chen, C.J.; Chen, C.M.; Chang, H.S.; Wu, Y.R. The use of evoked potentials for clinical correlation and surgical outcome in cervical spondylotic myelopathy with intramedullary high signal intensity on MRI. *J. Neurol. Neurosurg. Psychiatry* **2004**, *75*, 256–261.

21. Morishita, Y.; Hida, S.; Naito, M.; Matsushima, U. Evaluation of cervical spondylotic myelopathy using somatosensory-evoked potentials. *Int. Orthop.* **2005**, *29*, 343–346. [CrossRef] [PubMed]

22. Tessitore, E.; Broc, N.; Mekideche, A.; Seeck, M.; Truffert, A.; Vargas, M.I.; Schonauer, C.; Schaller, K. A modern multidisciplinary approach to patients suffering from cervical spondylotic myelopathy. *J. Neurosurg. Sci.* **2019**, *63*, 19–29. [CrossRef] [PubMed]

23. Keller, A.; von Ammon, K.; Klaiber, R.; Waespe, W. [Spondylogenic cervical myelopathy: Conservative and surgical therapy]. *Schweiz. Med. Wochenschr.* **1993**, *123*, 1682–1691. [PubMed]

24. White, A.A.; Panjabi, M.M. Update on the evaluation of instability of the lower cervical spine. *Instr. Course Lect.* **1987**, *36*, 513–520. [PubMed]

25. Hirabayashi, K.; Miyakawa, J.; Satomi, K.; Maruyama, T.; Wakano, K. Operative results and postoperative progression of ossification among patients with ossification of cervical posterior longitudinal ligament. *Spine* **1981**, *6*, 354–364. [CrossRef] [PubMed]

26. Chung, S.S.; Lee, C.S.; Chung, K.H. Factors affecting the surgical results of expansive laminoplasty for cervical spondylotic myelopathy. *Int. Orthop.* **2002**, *26*, 334–338. [CrossRef] [PubMed]

27. Alisauskiene, M.; Magistris, M.R.; Vaiciene, N.; Truffert, A. Electrophysiological evaluation of motor pathways to proximal lower limb muscles: A combined method and reference values. *Clin. Neurophysiol.* **2007**, *118*, 513–524. [CrossRef]

28. Fehlings, M.G.; Tetreault, L.A.; Riew, K.D.; Middleton, J.W.; Aarabi, B.; Arnold, P.M.; Brodke, D.S.; Burns, A.; Carette, S.; Chen, R.; et al. A Clinical Practice Guideline for the Management of Patients With Degenerative Cervical Myelopathy: Recommendations for Patients With Mild, Moderate, and Severe Disease and Nonmyelopathic Patients With Evidence of Cord Compression. *Global Spine J.* **2017**, *7*, 70S–83S. [CrossRef]

29. Hashizume, Y.; Iijima, S.; Kishimoto, H.; Yanagi, T. Pathology of spinal cord lesions caused by ossification of the posterior longitudinal ligament. *Acta Neuropathol.* **1984**, *63*, 123–130. [CrossRef]

30. Shi, R.; Pryor, J.D. Pathological changes of isolated spinal cord axons in response to mechanical stretch. *Neuroscience* **2002**, *110*, 765–777. [CrossRef]

31. Fehlings, M.G.; Wilson, J.R.; Kopjar, B.; Yoon, S.T.; Arnold, P.M.; Massicotte, E.M.; Vaccaro, A.R.; Brodke, D.S.; Shaffrey, C.I.; Smith, J.S.; et al. Efficacy and safety of surgical decompression in patients with cervical spondylotic myelopathy: Results of the AOSpine North America prospective multi-center study. *J. Bone Joint Surg. Am.* **2013**, *95*, 1651–1658. [CrossRef] [PubMed]

32. Mastronardi, L.; Elsawaf, A.; Roperto, R.; Bozzao, A.; Caroli, M.; Ferrante, M.; Ferrante, L. Prognostic relevance of the postoperative evolution of intramedullary spinal cord changes in signal intensity on magnetic resonance imaging after anterior decompression for cervical spondylotic myelopathy. *J. Neurosurg. Spine* **2007**, *7*, 615–622. [CrossRef] [PubMed]

33. Nakamura, M.; Fujiyoshi, K.; Tsuji, O.; Konomi, T.; Hosogane, N.; Watanabe, K.; Tsuji, T.; Ishii, K.; Momoshima, S.; Toyama, Y.; et al. Clinical significance of diffusion tensor tractography as a predictor of functional recovery after laminoplasty in patients with cervical compressive myelopathy. *J. Neurosurg. Spine* **2012**, *17*, 147–152. [CrossRef] [PubMed]

34. Wen, C.; Cui, J.L.; Liu, H.S.; Mak, K.C.; Cheung, W.Y.; Luk, K.D.K.; Hu, Y. Is diffusion anisotropy a biomarker for disease severity and surgical prognosis of cervical spondylotic myelopathy? *Radiology* **2014**, *270*, 197–204. [CrossRef] [PubMed]

35. Ebersold, M.J.; Pare, M.C.; Quast, L.M. Surgical treatment for cervical spondylitic myelopathy. *J. Neurosurg.* **1995**, *82*, 745–751. [CrossRef]

36. Yoon, S.T.; Hashimoto, R.E.; Raich, A.; Shaffrey, C.I.; Rhee, J.M.; Riew, K.D. Outcomes after laminoplasty compared with laminectomy and fusion in patients with cervical myelopathy: A systematic review. *Spine* **2013**, *38*, S183–S194. [CrossRef]

37. Maki, S.; Koda, M.; Kitamura, M.; Inada, T.; Kamiya, K.; Ota, M.; Iijima, Y.; Saito, J.; Masuda, Y.; Matsumoto, K.; et al. Diffusion tensor imaging can predict surgical outcomes of patients with cervical compression myelopathy. *Eur. Spine J.* **2017**, *26*, 2459–2466. [CrossRef]

38. Kim, H.-J.; Moon, S.-H.; Kim, H.-S.; Moon, E.-S.; Chun, H.-J.; Jung, M.; Lee, H.-M. Diabetes and smoking as prognostic factors after cervical laminoplasty. *J. Bone Joint Surg. Br.* **2008**, *90*, 1468–1472. [CrossRef]

39. Chen, Y.; Guo, Y.; Chen, D.; Wang, X.; Lu, X.; Yuan, W. Long-term outcome of laminectomy and instrumented fusion for cervical ossification of the posterior longitudinal ligament. *Int. Orthop.* **2009**, *33*, 1075–1080. [CrossRef]

40. Kawaguchi, Y.; Matsui, H.; Ishihara, H.; Gejo, R.; Yasuda, T. Surgical outcome of cervical expansive laminoplasty in patients with diabetes mellitus. *Spine* **2000**, *25*, 551–555. [CrossRef]

41. Tetreault, L.; Kopjar, B.; Côté, P.; Arnold, P.; Fehlings, M.G. A Clinical Prediction Rule for Functional Outcomes in Patients Undergoing Surgery for Degenerative Cervical Myelopathy: Analysis of an International Prospective Multicenter Data Set of 757 Subjects. *J. Bone Joint Surg. Am.* **2015**, *97*, 2038–2046. [CrossRef] [PubMed]

42. Okada, Y.; Ikata, T.; Yamada, H.; Sakamoto, R.; Katoh, S. Magnetic resonance imaging study on the results of surgery for cervical compression myelopathy. *Spine* **1993**, *18*, 2024–2029. [CrossRef] [PubMed]

43. Fukushima, T.; Ikata, T.; Taoka, Y.; Takata, S. Magnetic resonance imaging study on spinal cord plasticity in patients with cervical compression myelopathy. *Spine* **1991**, *16*, S534–S538. [CrossRef] [PubMed]

44. Morio, Y.; Teshima, R.; Nagashima, H.; Nawata, K.; Yamasaki, D.; Nanjo, Y. Correlation between operative outcomes of cervical compression myelopathy and mri of the spinal cord. *Spine* **2001**, *26*, 1238–1245. [CrossRef]

45. Rao, A.; Soliman, H.; Kaushal, M.; Motovylyak, O.; Vedantam, A.; Budde, M.D.; Schmit, B.; Wang, M.; Kurpad, S.N.; Motovylak, O. Diffusion Tensor Imaging in a Large Longitudinal Series of Patients With Cervical Spondylotic Myelopathy Correlated With Long-Term Functional Outcome. *Neurosurgery* **2018**, *83*, 753–760. [CrossRef]

46. Dong, F.; Wu, Y.; Song, P.; Qian, Y.; Wang, Y.; Xu, L.; Yin, M.; Zhang, R.; Tao, H.; Ge, P.; et al. A preliminary study of 3.0-T magnetic resonance diffusion tensor imaging in cervical spondylotic myelopathy. *Eur. Spine J.* **2018**, *27*, 1839–1845. [CrossRef]

Journal of
Clinical Medicine

Article

The Functional Relevance of Diffusion Tensor Imaging in Patients with Degenerative Cervical Myelopathy

Stefania d'Avanzo [1,*,†], Marco Ciavarro [2,†], Luigi Pavone [2], Gabriele Pasqua [3,4], Francesco Ricciardi [2], Marcello Bartolo [2], Domenico Solari [1], Teresa Somma [1], Oreste de Divitiis [1], Paolo Cappabianca [1] and Gualtiero Innocenzi [2]

[1]   Division of Neurosurgery, Department of Neurosciences, Reproductive and Odontostomatological Sciences, Federico II University of Naples, 80126 Naples, Italy; domenico.solari@unina.it (D.S.); teresa.somma85@gmail.com (T.S.); dedivitiis.oreste@gmail.com (O.d.D.); paolo.cappabianca@unina.it (P.C.)

[2]   I.R.C.C.S. Neuromed, 86077 Pozzilli, Italy; marcociavarro@gmail.com (M.C.); bioingegneria@neuromed.it (L.P.); fricciardi1@yahoo.it (F.R.); bartolonrx@gmail.com (M.B.); innocenzigualtiero@tiscali.it (G.I.)

[3]   Medicine and Health Science Department, University of Molise, 86100 Campobasso, Italy; ing.gabrielepasqua@gmail.com

[4]   Human Neuroscience Department, Sapienza University of Rome, 00185 Rome, Italy

*   Correspondence: stefania86nch@gmail.com

†   These authors have contributed equally to this work.

Received: 8 May 2020; Accepted: 8 June 2020; Published: 11 June 2020

Abstract: (1) Background: In addition to conventional magnetic resonance imaging (MRI), diffusion tensor imaging (DTI) has been investigated as a potential diagnostic and predictive tool for patients with degenerative cervical myelopathy (DCM). In this preliminary study, we evaluated the use of quantitative DTI in the clinical practice as a possible measure to correlate with upper limbs function. (2) Methods: A total of 11 patients were enrolled in this prospective observational study. Fractional anisotropy (FA) values was extracted from DTI data before and after surgery using a GE Signa 1.5 T MRI scanner. The Nine-Hole Peg Test and a digital dynamometer were used to measure dexterity and hand strength, respectively. (3) Results: We found a significant increase of FA values after surgery, in particular below the most compressed level ($p = 0.044$) as well as an improvement in postoperative dexterity and hand strength. Postoperative FA values moderately correlate with hand dexterity ($r = 0.4272$, $R_2 = 0.0735$, $p = 0.19$ for the right hand; $r = 0.2087$, $R_2 = 0.2265$, $p = 0.53$ for the left hand). (4) Conclusion: FA may be used as a marker of myelopathy and could represent a promising diagnostic value in patients affected by DCM. Surgical decompression can improve the clinical outcome of these patients, especially in terms of the control of finger-hand coordination and dexterity.

Keywords: diffusion tensor imaging (DTI); fractional anisotropy (FA); cervical MRI; degenerative cervical myelopathy (DCM); myelopathy hand

---

1. Introduction

Degenerative cervical myelopathy (DCM) is the most common non-traumatic spinal cord disorder in patients over 55 years old [1–4]: It is a progressive spinal cord disease characterized by degenerative changes of the bone, ligaments and intervertebral disc of cervical spine [5,6].

DCM comprises a wide set of clinical features, including neck pain, motor and sensory deficits and bladder dysfunction [7,8]. Furthermore, peculiar loss of strength and hand dexterity (the so-called "clumsy hand" or "myelopathy hand") is observed in patients with DCM. Ono et al. [8] first reported a loss of intensity of adduction and extension in the ulnar of two or three fingers and an inability to grip and release rapidly with these fingers. Furthermore, the occurrence of "myelopathy hand" has been demonstrated to be a crucial clinical sign to achieve an early suspicion of pyramidal tract damage [9–11].

Conventional T2-weighted magnetic resonance imaging (T2WI) is an integral part of DCM patient evaluation. There is some evidence from a largest prospective multicenter magnetic resonance imaging (MRI) study that signal changes have some relevance in terms of its correlation with baseline and outcome. T2WI shows an increased signal intensity in the compressed part of the spinal cord; however, this abnormal MR signal has low sensitivity for structural change of the cord in cervical myelopathy and it is not predictive of neurological function before and after surgical treatment [12–17].

Diffusion tensor imaging (DTI) is an advanced imaging technique that has been proposed to assess DCM-associated demyelination and axonal damage. DTI provides also quantitative information about the white matter microarchitecture. Fractional anisotropy (FA) is a quantitative DTI parameter that measures the tendency of water to spread in a preferred direction within a group of axons and it is a function of the axonal density and integrity of white matter fibers as well as the degree of myelination [18–20]. Normative values of FA for healthy subjects were found to be 0.68 ± 0.05, after correcting for age and sex. Hence, a decrease of FA value highlights fiber tracts impairment. Several studies showed significant decrease of FA values at the most compressed level, but also at distant areas [21,22].

DTI values, as compared to conventional MRI, are more sensitive in the detection of DCM patients, especially in the early stage of the disease; quantitative analysis of its parameters helps in the definition of myelopathy severity and can predict the outcomes of surgical treatments [23–26].

In this study, we aimed to define the diagnostic value of quantitative DTI in patients with DCM, measuring the correlation between the FA values and hand-motor performance, i.e., dexterity and hand strength—as measured via function test batteries, thus determining its functional relevance.

2. Experimental Section

2.1. Participants

This is a prospective observational study reporting preliminary data on 11 patients (6 females and 5 males; mean age 57.64 ± 10.47 years). All subjects included in the sample were diagnosed with degenerative cervical myelopathy, as well as with myelopathic hand. The severity of cervical myelopathy was assessed using the modified Japanese Orthopedic Association (mJOA) score [27]. Each patient underwent surgery via anterior cervical discectomy and fusion (ACDF) at the I.R.C.C.S. Neuromed di Pozzilli (Isernia, Italy). A 1.5 T MRI scan from 24 to 48 h prior to the surgery and another MRI scan 3 months after the surgery were acquired for each patient. In order to evaluate the severity of the clumsy hand, the measurement of strength and hand dexterity was performed the same day of the MRI scans. Patients with cerebral palsy, rheumatoid arthritis or other spinal diseases were excluded from this study. Protocol was approved by the Ethical Committee of the I.R.C.C.S. Neuromed (Ethical Approval Code: 11/17 21-12-2017).

2.2. Diffusion Tensor Imaging (DTI) Acquisition and Analysis

In order to study and quantify changes in white matter structural integrity, patients underwent a cervical MRI scan with Diffusion Tensor Imaging (DTI), with a focus at the pathological segment, 24–48 h before and three months after the surgery. A GE Signa 1.5 T MRI scanner was used to acquire MRI data. The MRI protocol included Structural 2D T2-weighted images (Slice Thickness 4 mm, Repetition Time 6700 ms, Echo Time 95.9 ms, Matrix Size 320 × 224, Field of View 220 mm, Flip Angle 90°) acquired both in the axial and sagittal plane, and DTI images with 16 diffusion directions (b = 1000 s/mm^2, 1b0, Repetition Time 10,000 ms, Echo Time 100 ms, Matrix Size 92 × 64). Image analysis was performed using the 3D Slicer software [28] and Fractional Anisotropy (FA) was extracted from DTI data [29,30]. Image processing pipeline comprised registration of anatomical T2WI with DTI images, using a 27 degrees of freedom BSpline registration. The accuracy of registration was visually assessed by a neuroradiologist (MB). Different Regions of Interest (ROIs), using the T2 images as reference, were created on a color-coded FA map in correspondence of the anatomical levels C2-C3, C3-C4, C4-C5, C5-C6 and C6-C7 to compute the aforementioned DTI parameters. The ROIs were designed by neuroradiologist and neurosurgeon (MB and GI), including both the white matter and the grey matter and excluding the cerebrospinal fluid (CSF), as described by Thurnher et al. [31].

2.3. Measurement of Dexterity and Hand Strength

The Nine-Hole Peg Test (NHPT) was used to assess the "digital dexterity" of the hand. Each subject performed "fine" grasping movement of nine pegs and released them in a wooden base. Once this phase was completed, the patient was instructed to remove each peg one by one, with the same hand. The test ended when all the pegs were placed inside the lid. The patients repeated the test twice for each hand and the average execution time was taken as result [32].

"Jamar" type digital dynamometer was used to measure the hand strength. The patient was instructed to sit down with the trunk in a neutral position, with the abducted shoulders aligned with each other on the frontal plane, the elbow flexed at 90°, the forearm in a neutral position, the wrist in extension between 0° and 20° and with an ulnar deviation between 0° and 15°. The dynamometer was supported by the operator's hand to prevent possible loss of strength by the patient. Using the same setting, for the dexterity assessment the patients performed two tests and then the average value was taken as result. The first rehearsal was conducted with the dominant limb (right for all the patients) [33] (Figure 1). All these tests were performed 24–48 h before and three months after surgery.

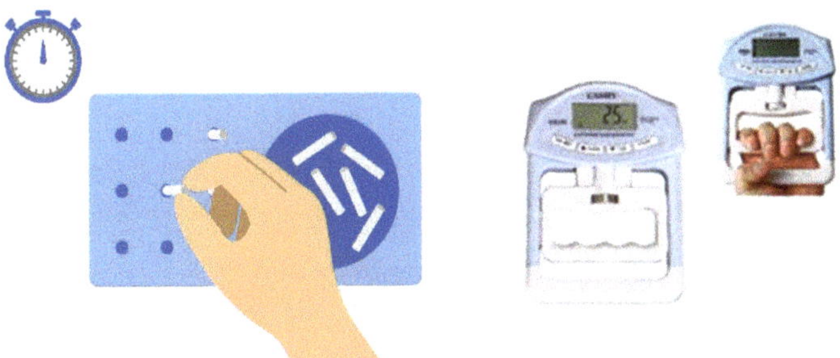

Figure 1. On the left: Nine-Hole Peg Test (NHPT). On the right: digital dynamometer (Camry EH101).

2.4. Statistical Analysis

Data were analyzed using a paired *t*-test to compare the preoperative and postoperative Fractional Anisotropy values, dexterity and hand strength. A Pearson correlation analysis was then performed to assess the correlation between the postoperative FA values, strength and hand dexterity. A *p*-value < 0.05 was considered to be statistically significant.

3. Results

3.1. Baseline Characteristics

All patients underwent anterior cervical discectomy and fusion (ACDF) at 1 or 2 cervical levels. Eight patients (72.7%) showed high signal intensity (HSI) on T2WI; the mean mJOA score was 13.27 ± 2.61, therefore moderate myelopathy has been diagnosed in most patients (Table 1).

Table 1. Demographic, MRI characteristics and surgical level.

Case	Gender	Surgical Level	Age	T2 Hyperintensity Signal	Most Compressed Level	mJOA Pre-op
1	F	2	57	+	C5-C6	13
2	M	2	66	+	C3-C4	12
3	F	1	75	-	C5-C6	11
4	M	2	55	-	C5-C6	9
5	F	2	56	+	C5-C6	16
6	F	1	59	+	C3-C4	12
7	M	2	38	+	C5-C6	11
8	M	1	56	+	C5-C6	14
9	M	1	67	+	C5-C6	14
10	F	2	62	-	C4-C5	17
11	F	1	43	-	C5-C6	17

F: female; M: male; mJOA: modified Japanese Orthopedic Association score.

3.2. Diffusion Tensor Imaging (DTI)

The preoperative FA values were pathological ($<0.68 \pm 0.05$ according to [20,21]) only at C6/C7 level, mostly below the frequently compromised level (Table 2). A significant increase in FA values after surgery was found at C5/C6 and C6/C7 level (paired *t*-test, $p = 0.005$ and $p = 0.002$ respectively) (Figure 2).

Table 2. Fractional anisotropy (FA) values of the anatomical level.

FA	C2/C3	C3/C4	C4/C5	C5/C6	C6/C7
PRE	0.72	0.74	0.70	0.67	0.61
POST	0.72	0.74	0.73	0.72	0.66

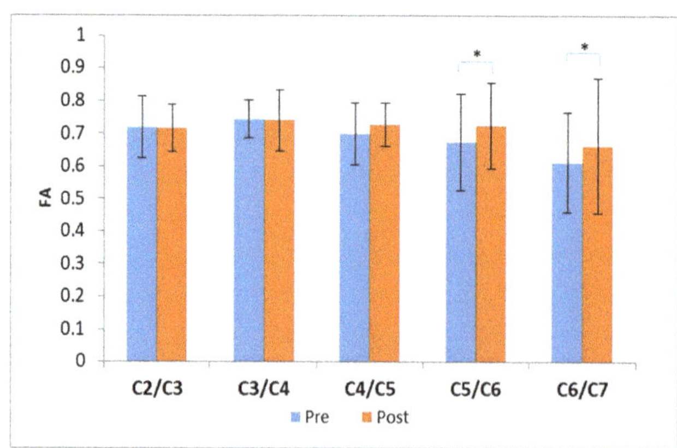

Figure 2. Fractional anisotropy (FA) values. Data are shown as mean values. The FA corresponding to the pre-surgery evaluation are shown in blue, whereas the FA corresponding to the post-surgery evaluation are shown in orange. Significant differences are shown with * ($p < 0.05$).

We computed the FA values in correspondence of the most compressed anatomical level, which was the target of the surgery, and also in the anatomical levels immediately above and below.

The preoperative FA value was pathological (Table 3) below the most compressed level; in fact, a statistically significant increase of FA postoperatively was observed at the lower level ($p = 0.044$). (Figure 3).

Table 3. FA values of the most compressed level (site of surgery) and the upper and lower level. The pathological FA value (<0.68 ± 0.05 [20]) is shown in red color.

| Case | Pre-Surgery (24–48 h) | | | Post-Surgery (Mean Follow-up 12 ± 2 Weeks) | | |
	Above FA	Most Compressed Level FA	Below FA	Above FA	Most Compressed Level FA	Below FA
1.	0.69	0.51	0.57	0.76	0.56	0.420
2.	0.63	0.76	0.70	0.60	0.71	0.66
3.	0.81	0.73	0.73	0.85	0.85	0.82
4.	0.70	0.79	0.72	0.71	0.84	0.82
5.	0.65	0.62	0.67	0.65	0.90	0.67
6.	0.82	0.73	0.73	0.72	0.52	0.60
7.	0.80	0.88	0.50	0.78	0.82	0.66
8.	0.63	0.59	0.63	0.72	0.59	0.89
9.	0.85	0.84	0.76	0.72	0.54	0.42
10.	0.77	0.54	0.40	0.78	0.74	0.59
11.	0.61	0.66	0.71	0.75	0.71	0.75
Mean	0.72 ± 0.08	0.69 ± 0.12	0.64 ± 0.11	0.72 ± 0.08	0.71 ± 0.15	0.70 ± 0.14

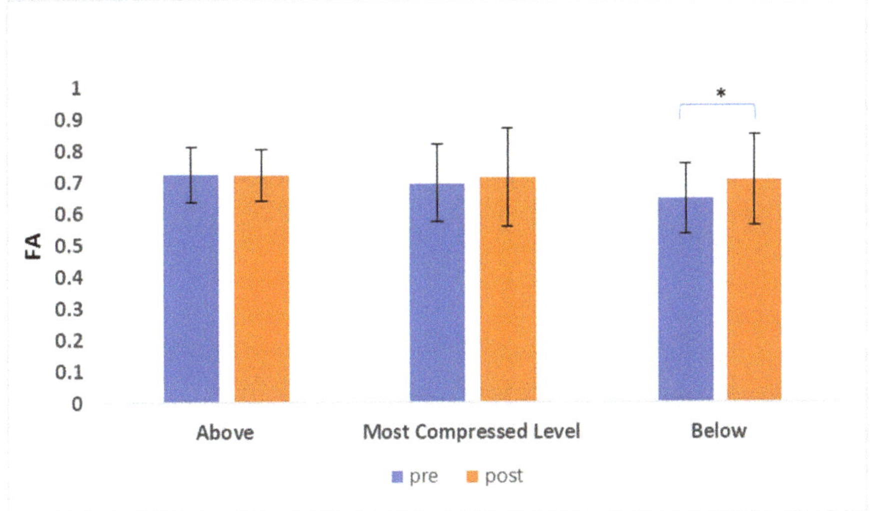

Figure 3. FA values of the most compressed level, the site of surgery, and the anatomical levels immediately above and below. Data are shown as mean values. The FA corresponding to the pre-surgery evaluation are shown in blue, whereas the FA corresponding to the post-surgery evaluation are shown in orange. Significant differences are shown with * ($p < 0.05$).

3.3. Dexterity and Hand Strength

During the preoperative and postoperative evaluations, the mean values of hand "dexterity" were 28.8 and 25.6 s for the right hand, respectively, and 29 and 24.5 s for the left hand, respectively (Table 4). There was a significant improvement of hand "dexterity" ($p = 0.002$) for the left hand with a 15.4% reduction in the time needed for completing the task, while for the right hand, the improvement was not statistically significant ($p = 0.057$) (Figure 4).

Furthermore, the mean values of right hand strength were 26.2 vs. 27.4 kg, while the mean values of left hand strength were 24.4 and 25.7 kg (Table 4); no significant improvement of the postoperative strength was found for both hands ($p = 0.055$ and 0.068) (Figure 5).

Table 4. Measurement of dexterity and hand strength. The percentage of improvements for NHPT and hand strength was reported in green.

| | Right Hand | | | | | | | | | | | | Left Hand | | | | | | | | | | | |
| | NHPT | | | Hand Strength | | | NHPT | | | Hand Strength | | |
	pre	post	%	pre	post	%	pre	post	%	pre	post	%
1	24.0	18.9	−21.3	23.4	25.6	9.4	26.4	22.0	−16.7	20.9	21.6	3.3
2	38.1	27.2	−28.6	31.3	35.8	14.4	33.1	24.5	−26.0	35.4	34.3	−3.1
3	32.4	34.0	5.1	15.1	18.2	20.6	39.2	36.0	−8.2	8.9	15.7	76.4
4	31.2	24.0	−23.0	18.5	16.2	−12.4	27.7	25.2	−9.0	20.1	16.5	−17.9
5	34.7	30.1	−13.3	14.1	17.4	23.4	47.4	37.0	−22.0	6.9	8.4	21.9
6	43.1	35.5	−17.6	12.0	13.5	13.0	28.8	22.1	−23.1	23.1	22.1	−4.3
7	20.0	19.3	−3.5	66.3	64.3	−3.1	21.3	19.1	−10.1	54.5	55.2	1.3
8	20.2	24.8	22.8	39.4	40.7	3.3	22.3	22.2	−0.7	34.0	36.0	5.9
9	23.1	31.2	35.1	29.3	28.4	−3.2	23.9	25.1	5.0	27.6	28.7	4.0
10	30.0	19.7	−34.5	18.0	19.5	8.3	30.5	20.0	−34.4	15.0	21.5	43.3
11	19.8	16.8	−14.9	20.9	21.5	2.9	18.1	16.5	−8.8	22.3	23.2	3.8
Mean	28.8	25.6	−11.1	26.2	27.4	4.4	29.0	24.5	−15.4	24.4	25.7	5.4
p-value	0.057			0.055			0.002			0.068		

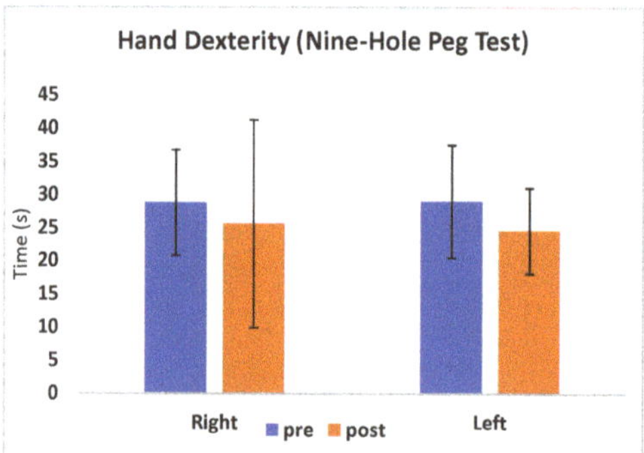

Figure 4. The average values of hand dexterity corresponding to the pre-surgery evaluation are shown in blue, whereas the values of hand dexterity corresponding to the post-surgery evaluation are shown in orange, distinguishing between right and left hand.

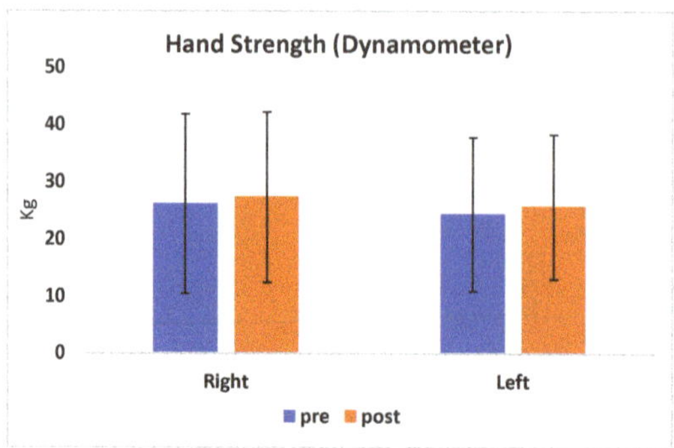

Figure 5. The average values of hand strength corresponding to the pre-surgery evaluation are shown in blue, whereas the values of hand strength corresponding to the post-surgery evaluation are shown in orange, distinguishing between right and left hand.

3.4. Correlation between the Postoperative FA Values and Strength and Hand Dexterity

A weak linear correlation between the postoperative FA values measured at lower surgical level and the dexterity scores was observed; indeed, the time to perform fine grasping task was inversely proportional to the FA value (Pearson coefficient r = 0.4272, coefficient of determination R_2 = 0.0735, *p*-value = 0.19 for the right hand; r = 0.2087, R_2 = 0.2265, *p*-value = 0.53 for the left hand) (Figure 6).

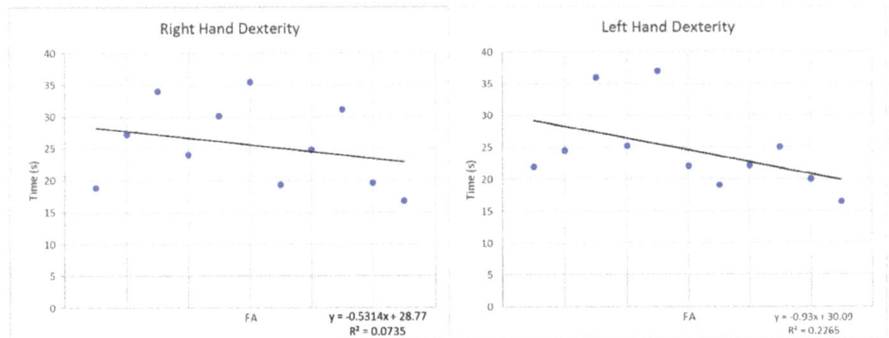

Figure 6. Correlation between the postoperative FA values and hand dexterity.

The FA values were also positively correlated with the postoperative hand strength data: There was a weak correlation between the two variables, so that the higher were the FA values, the higher was the strength of each hand (r = −0.0216, R_2 = 0.0068, *p*-value = 0.95 for the right hand; r = 0.0035 R_2 = 0.0376, *p*-value = 0.99 for the left hand) (Figure 7).

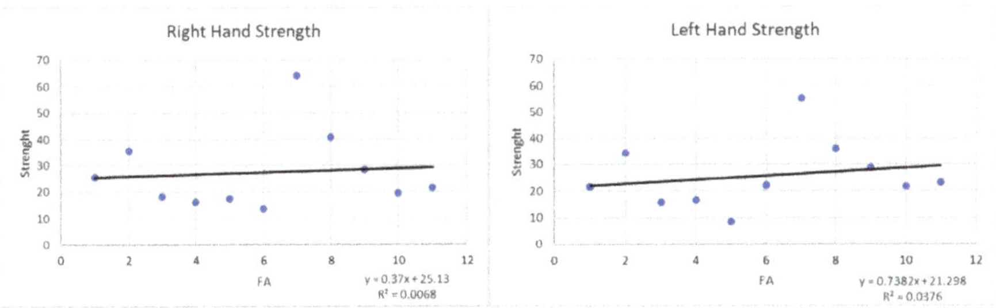

Figure 7. Correlation between the postoperative FA values and hand strength.

4. Discussion

Degenerative cervical myelopathy (DCM) is an insidiously progressive condition, usually showing a chronic course of clinical symptoms: impairment of gait, weakness, spasticity, clumsy hands, and sphincter disorders [1–7]. Conventional MR examination has played a key role in the diagnosis of cervical spondylosis, proving hypertrophy of the posterior longitudinal ligament and ligament flavum, cervical disc herniation, and cervical spinal stenosis. The compressed part of the spinal cord shows a specific high signal intensity (HSI) on T2WI. T2 HSI is often used to diagnose DCM, but this finding is not observed in each patient with clinical signs of myelopathy and its sensitivity is reported to be quite low (between 15% and 65%). Additionally, T2 HSI is generally observed only in the later stages of the disease [12–17]. A promising MR technique, diffusion tensor imaging (DTI), has been investigated for estimating the neural tissue integrity in spinal cord. As compared to conventional MR imaging, DTI parameters are more sensitive in detecting DCM, especially at the early stages [18–26] and they might represent a helpful tool for educating and monitoring subjects with asymptomatic spinal cord compression [34].

Unlike free water, water molecule diffusion in human is hindered by cell alignment pattern, cell membranes and other intracellular and extracellular structures showing anisotropy. Fractional anisotropy (FA) measures the tendency of water to spread in a preferred direction within a group of axons. It is a function of the axonal density and integrity of white matter fibers, as well as of their degree of myelination [35]. A decrease of FA corresponds to a damage of pyramidal tracts; in fact its value is significantly reduced in DCM patients, as compared to healthy subjects [20]. In our sample, the FA value was pathological (<0.68 ± 0.05 [21,22]) only below the most compressed level, i.e., below the segment approached surgically. It is also important to recognize that FA measures at the site of compression are sometimes difficult to obtain particularly in patients with considerable cord compression. We observed a significant increase in FA values in the postoperative course at the level just below the most compressed one, supporting that this parameter depicts a structural damage of the descending pyramidal pathways. This finding proves that DCM-associated demyelination and axonal damage afflicted both the myelopathic lesion and the distal sites over the chronic course of the disease [2].

Recent studies demonstrated a strong correlation between FA and specific clinical assessments, including mJOA score [17–26]. Shen et al. [23] showed that the mJOA score is a reasonable predictor of surgical outcome in DCM; nonetheless, a model inclusive of FA value provides superior predictive ability. Rajasekaran et al. claimed that a postoperative worsening of DTI indices is associated with a poor prognosis for neurological recovery [36]. Dong et al. [20] asserted that FA value of spinal cord was associated with postoperative recovery of spinal cord function and that DTI may play a significant role in diagnosing and predicting the development of DCM. The patients with severe DCM, who presented a higher FA value at the compressed level, were most likely to achieve a better functional recovery after decompression surgery [37]. This might identify FA as a potential positive predicting factor of postoperative outcomes: Therefore, DTI could be considered not only a complementary diagnostic analysis, but rather a crucial tool in order to identify the best candidates to surgery [38].

In particular, our study focused on one of the most common disorders in DCM patients: The myelopathy hand. Finger disability is a typical sign of degeneration of the corticospinal tracts and occurs only in patients with spinal cord lesions above C6-C7 level [9,10]. Doita et al. [39] showed a good correlation between the more severe cervical myelopathy and the loss of hand dexterity; Murphy et al. [40] demonstrated a strong correlation between the Nine-Hole Peg Test (NHPT) and FA values, showing that patients with moderate myelopathy performed the test in a longer time as compared to the control cases. The hand strength significantly differs between healthy subjects and myelopathic patients and its value is often influenced by age and sex.

Our patients show a more evident impairment of hand function in performing a precision grip, as assessed by the NHPT, a specific test which is regularly performed to evaluate manual dexterity in patients with multiple sclerosis and which was previously used to distinguish healthy subjects from patients with DCM. The hand "dexterity" was improved three months after surgery (11.1% for the right hand (Patient 1; 2; 4; 5; 6; 7; 10; 11), 15.4% for the left hand (Patient 1; 2; 3; 4; 5; 6; 7; 8; 10; 11)), with a moderate correlation between postoperative FA values and dexterity data; therefore, the time to carry out the test was reduced as FA values increase. The hand strength measured using a digital dynamometer showed a slight improvement at postoperative follow-up (4.4% for the right hand, 5.4% for the left hand); however, it was weakly associated with postoperative FA values.

In these patients, the damage of the corticospinal tracts determines a finger "spasticity", which is evident when the patient is asked to reopen the hand previously forcibly closed with abnormal prolongation of voluntary de-contracting. Spasticity represents a complex clinical sign that greatly compromises the hand dexterity and the ability in performing voluntary movements in myelopathic patients [41].

Our results confirmed that FA can be claimed as a marker of myelopathy presenting both diagnostic and potential prognostic value in patient affected by DCM, as depicting the functional status of the spinal cord. Indeed, surgical decompression can improve the clinical outcomes of these patients, especially in terms of control of "fine" grasping.

Nevertheless, our study has several limitations. First, the MRI scanner used is 1.5 T, which has a good enough resolution for DTI analysis, even though 3 T scanners can have better performance. The lack of a gold standard for the diagnostic imaging of DCM and the current high standard technical requirements for diffusion weighted imaging could represent the biases of this research. Finally, another limitation of the study is the relatively small sample size of our patients.

5. Conclusions

The diagnosis of degenerative cervical myelopathy includes a complex clinical picture: The patient's history, imaging and neurological status. The introduction of DTI allowed detecting spinal cord damage even at the earlier myelopathy stages, compared to the T2-weighted MR features. The combination of advanced imaging methods and diagnostic clinical tests for "Clumsy Hand" can help to accurately select the patients to be treated surgically and also to provide promising details in terms of predicting the outcomes. Further studies with larger case series and longer follow-up are needed to validate our results.

Author Contributions: Conceptualization: S.d., M.C.; G.I.; Data curation, F.R., M.B. and O.d.D.; formal analysis, F.R., M.B. and O.d.D.; methodology, L.P., G.P., F.R. and M.B.; software, L.P.; project administration, S.d., M.C., P.C. and G.I.; software, L.P., G.P.; supervision, D.S., T.S., P.C. and G.I.; validation, L.P. and G.P.; writing—original draft, S.d. and M.C.; writing—review and editing, S.d., M.C., L.P., G.P., D.S., T.S., P.C. and G.I. All authors have read and agree to the published version of the manuscript.

Funding: This research received no external funding.

Conflicts of Interest: The authors declare no conflicts of interest.

References

1. Wilson, J.R.; Tetreault, L.A.; Kim, J.; Shamji, M.F.; Harrop, J.S.; Mroz, T.; Cho, S.; Fehlings, M.G. State of the Art in Degenerative Cervical Myelopathy: An Update on Current Clinical Evidence. *Neurosurgery* **2017**, *80*, S33–S45. [CrossRef] [PubMed]

2. Karadimas, S.K.; Gatzounis, G.; Fehlings, M.G. Pathobiology of cervical spondylotic myelopathy. *Eur. Spine J.* **2015**, *24*, 132–138. [CrossRef] [PubMed]

3. Fehlings, M.G.; Tetreault, L. The aging of the global population: The changing epidemiology of disease and spinal disorders. *Neurosurgery* **2015**, *77*, 1–5. [CrossRef] [PubMed]

4. Nouri, A.; Tetreault, L. Degenerative cervical myelopathy: Epidemiology, genetics and pathogenesis. *Spine* **2015**, *40*, 675–693. [CrossRef] [PubMed]

5. Gurnam, V. Cervical Myelopathy: Pathophysiology, Diagnosis, and Management. *Spine Res.* **2017**, *3*, 2–12.

6. Fujiyoshi, T.; Yamazaki, M. Static versus dynamic factors for the developmentof myelopathy in patients with cervical ossification of the posterior longitudinal ligament. *J. Clin. Neurosci.* **2010**, *17*, 320–324. [CrossRef]

7. Fujiyoshi, T.; Yamazaki, M.; Okawa, A.; Kawabe, J.; Hayashi, K.; Endo, T.; Cho, S.; Fehlings, M.G. A Clinical Practice Guideline for the Management of Degenerative Cervical Myelopathy: Introduction, Rationale and Scope. *Glob. Spine J.* **2017**, *7*, 21–27.

8. Alian, J.; Micev, M.D. Cervical Radiculophaty and Myelopathy: Presentations in the Hand. *J. Hand Am.* **2013**, *38*, 2478–2481.

9. Ono, K.; Ebara, S.; Fuji, T.A.K.E.S.H.I.; Yonenobu, K.A.Z.U.O.; Fujiwara, K.E.I.J.U.; Yamashita, K.A.Z.U.O. Myelopathy hand: New clinical signs of cervical cord damage. *J. Bone Jt. Surg. Br.* **1987**, *69*, 215–219. [CrossRef]

10. Hosono, N.; Makino, T.; Sakaura, H.; Mukai, Y.; Fuji, T.; Yoshikawa, H. Myelopathy hand: New evidence of the classical sign. *Spine* **2010**, *35*, 273–277. [CrossRef] [PubMed]

11. Yukawa, Y.; Nakashima, H. Quantifiable tests for cervical myelopathy; 10-s grip and release test and 10-s step test: Standard values and aging variation from healthy volunteers. *J. Orthop. Sci.* **2013**, *18*, 509–513. [CrossRef]

12. Nouri, A.; Martin, A.R. Magnetic resonance imaging assessment of degenerative cervical myelopathy: A review of structural changes and measurement techniques. *Neurosurg. Focus* **2016**, *40*, 5–12. [CrossRef] [PubMed]

13. Cowley, P. Neuroimaging of Spinal Canal Stenosis. *Magn. Reson. Imaging Clin. N. Am.* **2016**, *24*, 523–539. [CrossRef] [PubMed]

14. Tetreault, L.A.; Dettori, J.R. Systematic review of magnetic resonance imaging characteristics that affect treatment decision making and predict clinical outcome in patients with cervical spondylotic myelopathy. *Spine* **2013**, *38*, 89–110. [CrossRef] [PubMed]

15. Karpova, A.; Arun, R. Do quantitative magnetic resonance imaging parameters correlate with the clinical presentation and functional outcomes after surgery in cervical spondylotic myelopathy? A prospective multicenter study. *Spine* **2014**, *39*, 1488–1497. [CrossRef] [PubMed]

16. Li, F.; Chen, Z. A meta-analysis showing that high signal intensity on T2-weighted MRI is associated with poor prognosis for patients with cervical spondylotic myelopathy. *J. Clin. Neurosci.* **2011**, *18*, 1592–1595. [CrossRef] [PubMed]

17. Nouri, A.; Martin, A.R. The relationship between MRI signal intensity changes, clinical presentation, and surgical outcome in degenerative cervical myelopathy: Analysis of a global cohort. *Spine* **2017**, *42*, 1851–1858. [CrossRef]

18. Guan, X.; Fan, G. Diffusion tensor imaging studies of cervical spondylotic myelophaty: A systemic review and meta-analysis. *PLoS ONE* **2015**, *10*, 1–12. [CrossRef]

19. Song, T.; Chen, V.J. Diffusion tensor imaging in the cervical spinal cord. *Eur. Spine J.* **2011**, *20*, 422–428. [CrossRef]

20. Dong, F.; Wu, F. A preliminary study of 3.0-T magnetic resonance diffusion tensor imaging in cervical spondylotic myelopathy. *Eur. Spine J.* **2018**, *27*, 1839–1845. [CrossRef]

21. Chagawa, K.; Nishijima, S. Normal values of diffusion tensor magnetic resonance imaging parameters in the cervical spinal cord. *Asian Spine J.* **2015**, *9*, 541–547. [CrossRef] [PubMed]

22. Wei, L.F.; Wang, S.S. Analysis of the diffusion tensor imaging parameters of a normal cervical spinal cord in healthy population. *J. Spinal Cord Med.* **2017**, *40*, 338–345. [CrossRef] [PubMed]

23. Shen, C.; Xu, H. Value of conventional MRI and DTI parameters in predicting surgical outcome in patients with DCM. *J. Back Musculoskelet. Rehabil.* **2018**, *31*, 525–532. [CrossRef]

24. Wang, K.; Chen, Z. Evaluation of DTI parameter ratio and diffusion tensor tractography grading in the diagnosis and prognosis prediction of cervical spondylotic myelopathy. *Spine* **2017**, *42*, 202–210. [CrossRef] [PubMed]

25. Yang, Y.M.; Yoo, W.K. The functional relevance of diffusion tensor imaging in comparison to conventional MRI in patients with cervical compressive myelopathy. *Skeletal Radiol.* **2017**. [CrossRef] [PubMed]

26. Landi, A.; Innocenzi, G. Diagnostic potential of the diffusion tensor tractography with fractional anisotropy in the diagnosis and treatment of cervical spondylotic and posttraumatic myelopathy. *Surg. Neurol. Int.* **2016**, *7*, 705–707. [CrossRef]

27. Tetreault, L.; Kopjar, B. The modified Japanese Orthopaedic Association scale: Establishing criteria for mild, moderate and severe impairment in patients with degenerative cervical myelopathy. *Eur. Spine J.* **2017**, *26*, 78–84. [CrossRef]

28. Fedorov, A.; Beichel, R. 3D Slicer as an Image Computing Platform for the Quantitative Imaging Network. *Magn. Reson. Imaging* **2012**, *30*, 1323–1341. [CrossRef]

29. Basser, P.J.; Mattiello, J.; LeBihan, D. MR diffusion tensor spectroscopy and imaging. *Biophys. J.* **1994**, *66*, 259–267. [CrossRef]

30. Westin, C.F.; Maier, S.E. Processing and visualization for diffusion tensor MRI. *Med. Image Anal.* **2002**, *6*, 93–108. [CrossRef] 965

31. Thurnher, M.M.; Mueller, M.C. Diffusion tensor MR imaging (DTI) metrics in the cervical spinal cord in asymptomatic HIV-positive patients. *Neuroradiology* **2011**, *53*, 585–592.

32. Olindo, S.; Signate, A. Quantitative assessment of hand disability by the Nine-Hole-Peg test (9-HPT) in cervical spondyloticmyelophaty. *J. Neurol. Neurosurg. Psychiatry* **2008**, *79*, 965–967.

33. Silberberg, N.; Kellor, M. Hand strenght and dexterity. *Am. J. Occup. Ther.* **1971**, *25*, 77–83.

34. Martin, A.R.; De Leener, B.; Cohen-Adad, J.; Cadotte, D.W.; Nouri, A.; Wilson, J.R.; Tetreault, L.; Crawley, A.P.; Mikulis, D.J.; Ginsberg, H.; et al. Can microstructural MRI detect subclinical tissue injury in subjects with asymptomatic cervical spinal cord compression? A prospective cohort study. *BMJ Open* **2018**, *8*, e019809. [CrossRef] [PubMed]

35. Clark, C.A.; Barker, G.J. Magnetic resonance diffusion imaging of the human cervical spinal cord in vivo. *Magn. Reson. Med.* **1999**, *41*, 1269–1273. [CrossRef]

36. Rajasekaran, S.; Kanna, R. Efficacy of Diffusion Tensor Imaging indices in assessing postoperative neural recovery in cervical spondylotic myelopathy. *Spine* **2016**, *42*, 8–13. [CrossRef] [PubMed]

37. Jones, J.G.; Cen, S.Y. Diffusion tensor imaging correlates with the clinical assessment of disease severity in cervical spondilotic myelopathy and predicts outcome following surgery. *AJNR Am. J. Neuroradiol.* **2013**, *34*, 471–478. [CrossRef] [PubMed]

38. Severino, R.; Nouri, A.; Tessitore, E. Degenerative cervical myelopathy: How to Identify the best responders to surgery? *J. Clin. Med.* **2020**, *9*, 759. [CrossRef]

39. Doita, M.; Sakai, H. Evaluation of impairment of hand function in patients with cervical myelophaty. *J. Spinal Disord.* **2006**, *19*, 276–280. [CrossRef]

40. Murphy, R.K.; Sun, P. Fractional anisotropy to quantify cervical spondylotic myelophaty severity. *J. Neurosurg. Sci.* **2018**, *62*, 406–412.

41. Smith, Z.A.; Barry, A.J.; Paliwal, M.; Hopkins, B.S.; Cantrell, D.; Dhaher, Y. Assessing hand dysfunction in cervical spondylotic myelopathy. *PLoS ONE* **2019**, *14*, e0223009. [CrossRef] [PubMed]

Article

Analysis of Cervical Spine Alignment and its Relationship with Other Spinopelvic Parameters after Laminoplasty in Patients with Degenerative Cervical Myelopathy

Seok Woo Kim [1,2,*], **Seung Bo Jang** [1], **Hyung Min Lee** [1], **Jeong Hwan Lee** [1], **Min Uk Lee** [1], **Jeong Woo Kim** [1] and **Jae Sung Yee** [1]

[1] Spine Center, Hallym University Sacred Heart Hospital, Hallym University, 896 Pyeongchon-dong, Dongan-gu, Anyang-si, Gyeonggi-do 431-070, Korea; ganzi404@naver.com (S.B.J.); lhm1011s@naver.com (H.M.L.); ljhsteve2338@gmail.com (J.H.L.); lmo932@hallym.or.kr (M.U.L.); hallymjw@naver.com (J.W.K.); id0917@naver.com (J.S.Y.)

[2] Department of Orthopaedic Surgery, Hallym University Sacred Heart Hospital, Hallym University, 896 Pyeongchon-dong, Dongan-gu, Anyang-si, Gyeonggi-do 431-070, Korea

* Correspondence: spinedrk@gmail.com; Tel.: +82-31-380-6000; Fax: +82-31-380-6008

Received: 24 January 2020; Accepted: 3 March 2020; Published: 5 March 2020

Abstract: For patients with kyphosis of the cervical spine, laminoplasty is usually incapable of improving neurological symptoms as it worsens kyphotic alignment. Thus, laminoplasty is not recommended in the presence of kyphotic alignment. Nevertheless, laminoplasty may be selected for myelopathy due to multiple-segment intervertebral disc herniation or ossification of posterior longitudinal ligament despite kyphotic alignment. This study examined whether cervical alignment influences surgical outcomes. Cervical alignment before the surgery was classified into lordosis and non-lordosis, and the non-lordosis group was subclassified into reducible and non-reducible groups to determine the change in cervical alignment before and after the surgery and to analyze its relationship with spinopelvic parameters. The lordosis group showed an increase in upper cervical motion (C0-2 Range of Motion (ROM), C0-2ROM/C0-7ROM) after surgery, while the non-lordosis group exhibited a decrease in C2-7ROM and C0-7ROM. The C0-2ROM was maintained without any reduction in the reducible group, while there was no significant change in cervical alignment and ROM of the non-reducible group. None of these changes showed significant association with the spinopelvic parameters of other sites. However, having a non-reducible type non-lordosis is not a proper indication for laminoplasty, as it does not change the alignment after surgery. Therefore, cervical alignment and reducibility should be identified before surgery.

Keywords: cervical alignment; kyphosis; spinopelvic parameter; laminoplasty; myelopathy

---

1. Introduction

Laminoplasty is an effective surgical method that is commonly applied in patients with myelopathy due to multiple-segment ossification of the posterior longitudinal ligament, cervical herniated intervertebral disc, and spinal stenosis. This method is known to produce desired results to recover the blood flow of the decompressed spinal cord through nerve decompression via effective posterior migration of spinal cord after surgery if the lordotic alignment of the cervical spine is well maintained. On the other hand, the kyphotic curvature of cervical spine is not usually recommended as a good indication for laminoplasty in the group of patients who have kyphotic alignment of cervical spine because laminoplasty cannot achieve effective posterior migration of spinal cord after surgery, and it possibly worsens the kyphotic alignment [1–7]. However, laminoplasty may be inevitably selected for myelopathy due to multiple-segment intervertebral disc herniation or ossification of posterior longitudinal ligament despite kyphotic alignment.

According to the report [8], when radiological changes such as clinical outcomes, sagittal alignment, and overall range of motion (ROM) were compared after laminoplasty was conducted for the group of patients who had cervical kyphosis with Cobb angle <10°, there were no significant statistical differences from the cervical lordosis group. Furthermore, when multiple segments are involved for patients having kyphotic alignment of cervical spine, laminoplasty showed the same surgical outcome as for the patients with cervical lordosis within the range of Cobb angle <10°.

A recent study has examined the cervical alignment of 958 normal asymptomatic individuals, finding that the ratio of the subjects with kyphotic alignment of cervical spine comprises 26.3%. The group with kyphotic alignment can be subdivided into reducible kyphosis, in which the kyphotic curvature recovers to lordotic alignment in the lateral radiograph upon extension, and non-reducible kyphosis, in which the kyphotic alignment is sustained upon extension. According to this classification, among the 26.3% of subjects with kyphotic alignment, 15.7% (of the full 958-person cohort) was determined as non-reducible kyphosis [9].

In this regard, the authors classified the preoperative cervical alignment of the patients who underwent laminoplasty in our clinic into lordosis and non-lordosis (reducible vs. non-reducible) to analyze the relationship between the change in cervical alignment before and after surgery and its relationship with other spinopelvic parameters. The authors extend the analysis in determining whether the preoperative cervical alignment affects postoperative cervical alignment and other spinopelvic parameters, and, particularly, whether the group with preoperative kyphotic alignment of cervical spine shows a significant change in postoperative cervical alignment and other spinopelvic parameters. This study intends to confirm whether kyphotic alignment of cervical spine would be a proper indication for laminoplasty in terms of radiological analysis.

2. Materials and Methods

2.1. Research Subjects

This study conducted a retrospective analysis by prospectively collecting data from 83 patients among 111 patients who underwent midline splitting double-door (French door) type laminoplasty based on a diagnosis of degenerative cervical myelopathy (DCM) from September 2008 to August 2015, excluding 28 patients with diagnosis of revision surgery, anterior cervical discectomy and fusion (ACDF) cases, total disc replacement (TDR) cases, infection, tumor, rheumatoid arthritis, etc. (Figure 1).

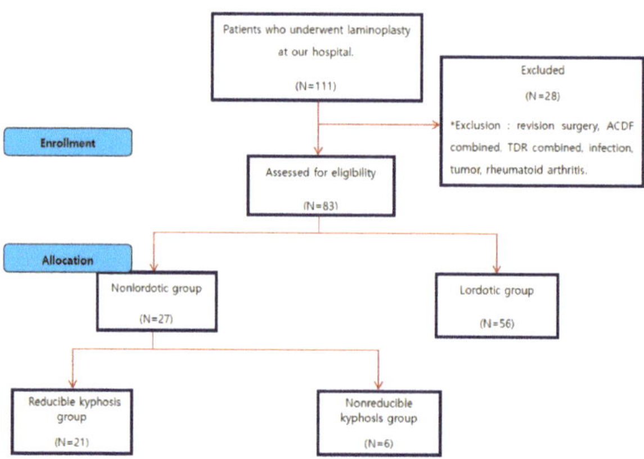

Figure 1. Flowchart of the study participants.

2.2. Surgical Technique

A posterior incision was made along the nuchal ligament to the line of the spinous processes. The semispinalis cervicis was partially detached from the lower margin of the C2 spinous process. Cervical laminae were exposed laterally to the medial aspect of the facet joints, and the interspinous ligaments were removed. The involved spinous processes were split sagittally with a 0.54 mm diameter Tomita saw (T-saw; Medtronics, Memphis, TN, USA). It cut along the midline epidural space in a caudal-to-cranial direction. The advancing tip of the polyethylene sleeve was grasped when it appeared in the flavectomy at the other end of the decompression zone. The T-saw was advanced through the sleeve so that it could be held securely, whereas the sleeve was withdrawn retrograde over the saw. At this point, the unsheathed saw spanned the midline of the spinal canal along the area to be decompressed. Each end was grasped with a special clamp or needle holder. The T-saw was pulled tight before initiating a reciprocating motion. The T-saw should fit snugly just at the midline of the inner wall of the laminar arch. Continuous reciprocating motion cut the midline of the inner wall of the laminar arch and the spinous processes in a ventral-to-dorsal direction away from the dura and spinal cord. The supra- and interspinous ligaments were automatically dissected at the midline. The saw was frequently lubricated with sterile saline solution to avoid excessive heat and friction [10]. After bilateral gutters for the hinges were carefully made with a high-speed burr at the transitional area between the facet joint and the laminae, spinal canal enlargement was achieved by opening the split laminae bilaterally with a spreader and placing allo-bone graft (Laminar Spacer-K; CG Bio, Seoul, Korea; Figure 2).

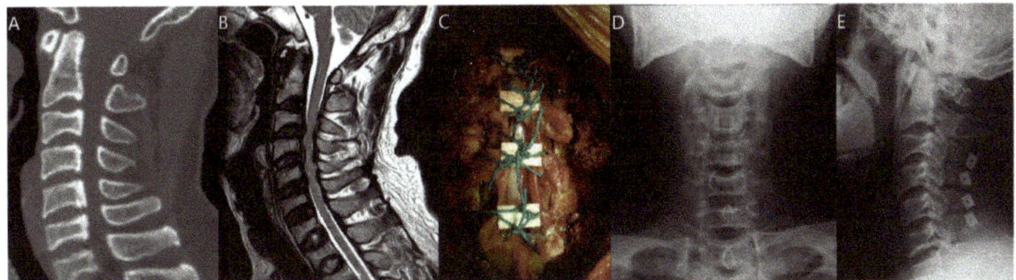

Figure 2. Pre-operative CT (**A**), pre-operative T2-weighted MRI (**B**), intra-operative photo (**C**), and post-operative AP (**D**) and lateral X-ray (**E**) of a 49-year-old male patient who underwent C3-6 midline splitting double-door type (French door type) laminoplasty.

2.3. Radiographic Measurement

This study used C-spine lateral X-rays, dynamic flexion–extension lateral radiography, and whole-spine lateral X-rays during the follow-up period before and after each surgery.

The radiographic protocol was standardized as follows. For each subject, cervical spine lateral radiographs were obtained with a 10 × 12 inch cassette at a 72 inch (182 cm) distance with the radiographic tube centered at the C4–C5 disc space with no magnification. The subjects were instructed to stand in a position with eyes looking forward and arms extended over their chests. Immediately after taking the cervical lateral neutral radiographs, flexion and extension views were obtained with a maximal neck flexion and extension position.

Whole-spine lateral radiographs were taken by using two 14 × 17 inch pieces of film in one 14 × 36 inch cassette at a 98.4 inch (250 cm) distance with the tube centered at the xiphoid process with no magnification. The subjects were instructed to stand in a position with eyes looking forward and arms crossed over their chests. The digital X-ray images were obtained and measured on a picture archiving and communication system (PiView, Infinitt, Seoul, Korea).

2.4. Radiological Analysis

C spine neutral X-ray was used to categorize the cervical alignment into lordotic and non-lordotic groups according to the Toyama classification [11], and the straight, sigmoid, and kyphosis groups were defined as a non-lordotic group (Figure 3). Afterwards, the analysis was conducted after subdividing the non-lordotic group into reducible non-lordotic group and non-reducible non-lordotic group, depending on the recovery of lordotic curve in the extension lateral X-ray view (Figure 4).

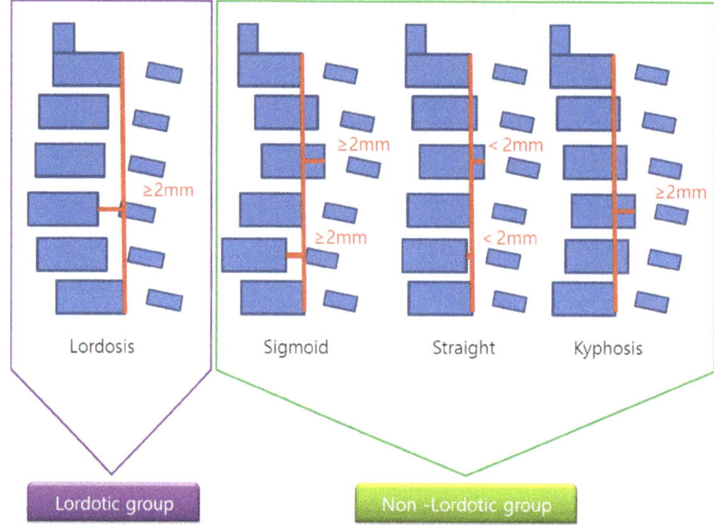

Figure 3. Diagram of lordosis and non-lordosis (kyphosis) groups.

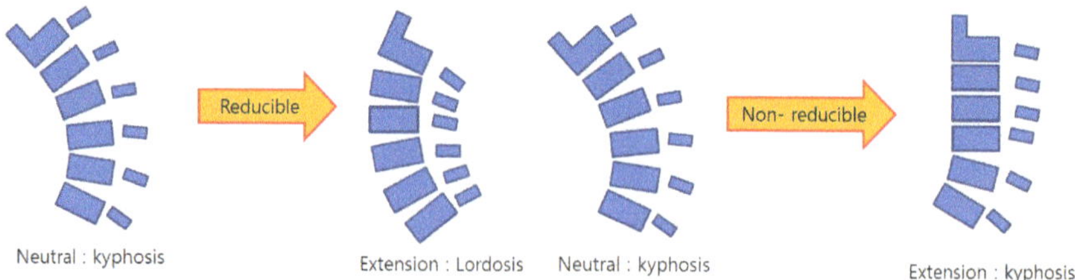

Figure 4. Diagram of reducible non-lordosis and non-reducible non-lordosis. Reducible, non-lordosis switches to lordosis in extension; non-reducible, non-lordosis (or kyphosis) maintains a non-lordotic (or kyphotic) alignment in extension.

In each group, C2-7 angle, C0-2 angle, C0-7 angle, T1 slope, C0-2 ROM, C0-7 ROM, and C2-7 ROM were measured in the neutral, flexion, and extension views.

This study also evaluated spinopelvic parameters including TK (thoracic kyphosis), LL (lumbar lordosis), SS (sacral slope), PT (pelvic tilt), and PI (pelvic incidence) to identify the relationship between preoperative and postoperative cervical alignment and thoracolumbosacral alignment. The definitions of each of measurements are explained in Table 1.

Table 1. Definitions of measurements.

Parameters	Definition
C2-7 Cobb angle	The intersection angle between the line perpendicular to the line parallel to the C2 lower endplate and the line perpendicular to the line parallel to the C7 lower endplate
C0-2 Cobb angle	The occipito-cervical angle, which is the intersection angle between the McGregor line and the line parallel to the C2 lower endplate and is used to evaluate the curvature of the upper cervical spine
C0-7 Cobb angle	The intersection angle between the McGregor line and the line parallel to the C7 lower endplate
C0-2/C0-7	The value of C0-2 Cobb angle divided by C0-7 Cobb angle
C2-7/C0-7	The value of C2-7 Cobb angle divided by C0-7 Cobb angle
T1 slope	The intersection angle between the tangent line and the upper plate of the T1 vertebral body
SS (Sacral Slope)	The angle formed by a line drawn along the endplate of the sacrum and a horizontal reference line
PT (Pelvic Tilt)	The angle formed by a line drawn from the midpoint of the sacral endplate to the center of the bicoxofemoral axis and a vertical and a vertical plumb line
PI (Pelvic incidence)	The angle formed by two vectors: (1) The line joining the bicoxo-femoral axis to the center of the sacral end plate and (2) A line perpendicular to the sacral endplate

2.5. Statistical Analysis

A *t*-test was used to analyze the measured parameters between the two groups, and ANOVA was utilized to analyze individual parameters among the three groups. Pearson correlation analysis was conducted to determine the correlation of the parameters among the groups. This study used the IBM SPSS software (version 22.0.0.1, IBM Corp., 2013, Armonk, NY, USA) in its statistical analysis. The statistical significance threshold was $p < 0.05$.

3. Results

The average age of all patients (59 males and 24 females) was 62.8, and the average follow-up period was 36.8 months. The lordosis group and the non-lordosis group included 56 (67.4%) and 27 (32.6%) patients, respectively; male to female ratios of 40:16 and 19:8, respectively; and average ages of 61.9 and 64.8, respectively (Table 2).

Table 2. Demographic data of the study participants.

Patient Who Underwent Laminoplasty (*n* = 83).			
	Lordosis	Non-Lordosis	*p*-Value *
No. of participants	56 (67.4%)	27 (32.6%)	
Sex ratio (M:F)	40:16	19:8	0.921
Age [95% CI]	61.9 [60.4~63.4]	64.8 [62.6~67.0]	0.292

CI, confidence interval. Chi-square test for sex ratio, independent t test for age. * Statistical significance ($p < 0.05$).

3.1. Post Operative Change of Curvature

The lordosis was maintained after surgery in 44 of the patients (78%) belonging to the lordosis group (N = 56) before the surgery. Although the lordosis was reduced in 12 patients (22%) after the surgery, only one patient exhibited no recovery of lordosis upon extension.

Of the patients belonging to the non-lordosis group (N = 27) before surgery, 21 (78%) fell under the reducible non-lordosis group showing the recovery of lordosis upon extension, and the remaining six (22%) fell under the non-reducible non-lordosis group exhibiting no recovery of lordosis upon extension. Eight patients (38%) in the reducible non-lordosis group ($N = 21$) changed into the lordosis group after the surgery, and seven (33%) and six (29%) shifted into the reducible non-lordosis group and the non-reducible non-lordosis group, respectively. All six patients belonging to the non-reducible non-lordosis group ($N = 6$) before surgery remained in the same group after the surgery (Figure 5).

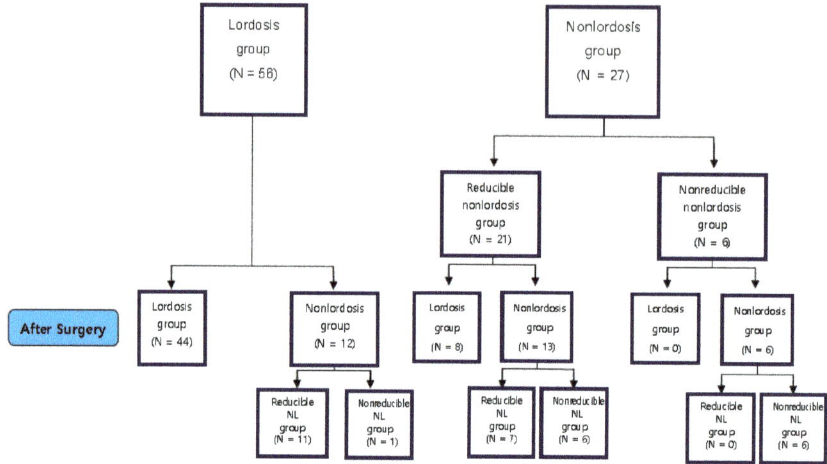

Figure 5. Post-operative change of curvature.

3.1.1. Comparison of Pre-Operative Radiological Parameters between the Lordosis and Non-Lordosis Groups

Among the parameters between the two groups, there were statistically significant differences in C0-2 angle ($p < 0.01$), C2-7 angle ($p < 0.01$), and the ratio of C0-2 ROM to C0-7 ROM ($p = 0.05$; Table 3). The reason why the ratio of C0-2 ROM to C0-7ROM in the non-lordosis group was high is that the more frequent use of upper cervical motion relatively compensates for the motion of the c-spine (C2-7 ROM). On the other hand, there were no statistically significant differences in other cervical and spinopelvic parameters (Table 3).

Table 3. Comparison of pre-operative radiographic parameters between lordosis and non-lordosis.

Parameter	Lordosis Mean (Degrees)	Non-Lordosis Mean (Degrees)	Between-Group Difference Mean (Degrees) (95% CI)	p-Value
C2-7angle	13.2(10.9~15.9)	1.1(−2.6~4.9)	12.1(7.7~16.5)	0.01 *
C0-2angle	24.2(21.2~26.8)	31.9(27.9~35.9)	−7.7 (−12.6~−2.8)	0.01 *
C0-7angle	36.8(33.3~40.4)	33.4(29.6~37.3)	3.4 (−2.4~9.1)	0.25
C0-2ROM	17.8(15.3~20.1)	23.1(16.2~30.1)	−5.3(−11.1~0.5)	0.07
C2-7ROM	34.3(30.7~37.6)	29.7(24.9~34.5)	4.6(−1.3~10.5)	0.12
C0-7ROM	52.4(47.7~56.4)	50.1(44.9~55.3)	2.3(−4.9~9.5)	0.53
C0-2/C0-7	0.3(0.3~0.4)	0.5(0.3~0.7)	−0.1(−0.3~0)	0.05 *
C2-7/C0-7	0.7(0.6~0.7)	0.6(0.5~0.7)	0.1(0~0.2)	0.18
T1slope	29.7(27.2~32.4)	26.9(23.5~30.3)	2.8(−1.6~7.2)	0.21
Lumbar lordosis	37.6(25.9~49.3)	26.4(12.3~39.9)	11.2(−0.8~23.3)	0.07
sacral slope	34.0(29.6~38.5)	30.9(21.7~40.1)	3.1(−5.9~12.1)	0.49
pelvic tilt	15.5(12.1~18.8)	16.5(9.6~23.4)	−1.0(−7.8~5.8)	0.76
pelvic incidence	47.4(40.8~53.9)	41.6 (23.8~50.5)	5.7 (−6.6~18.1)	0.35

CI, confidence interval; ROM, range of motion * Statistical significance ($p < 0.05$).

3.1.2. Comparison of Post-Operative Radiological Parameters between the Lordosis and Non-Lordosis Groups

The overall cervical spine ROM, including C2-7 ROM and C0-7 ROM, decreased after surgery, while C0-2ROM increased, though not statistically significantly. The statistically significant differences among the parameters between the two groups were observed in C2-7 angle ($p < 0.01$), C0-2 angle ($p = 0.02$), C0-7 angle ($p = 0.02$), the ratio of C0-2 ROM to C0-7 ROM (C0-2/C0-7 ROM; $p = 0.003$), and T1 slope ($p < 0.01$). The reason why the ratio of C0-2 ROM to C0-7 ROM was high in the non-lordosis group is that the relative compensation of the upper cervical motion (C0-2 ROM) for the motion of C-spine (C2-7 ROM) is maintained after the surgery (Table 4).

Table 4. Comparison of post-operative radiographic parameters between lordosis and non-lordosis.

Parameter	Lordosis Mean (Degrees)	Non-Lordosis Mean (Degrees)	Between-Group Difference Mean (Degrees) (95% CI)	*p*-Value
C2-7angle	11.3(8.4~14.4)	−0.1(-4.9~4.8)	11.4(5.9~16.9)	0.01 *
C0-2angle	27.2(24.2~30.5)	33.7(29.8~37.7)	−6.5(−11.8~-1.3)	0.02 *
C0-7angle	39.7(36.1~43.7)	32.9(29.5~36.3)	6.9(1.0~12.8)	0.02 *
C0-2ROM	21.7(18.6~24.7)	24.8(20.9~28.7)	−3.1(−8.2~2.0)	0.23
C2-7ROM	20.9(18.1~23.9)	19.8(14.6~24.9)	1.2(−4.2~6.6)	0.66
C0-7ROM	42.8(39.2~46.6)	40.9(35.2~46.7))	1.9(−4.7~8.5)	0.57
C0-2/C0-7	0.5(0.4~0.6)	0.6(0.5~0.7)	−0.1(−0.2~0)	0.03 *
C2-7/C0-7	0.5(0.4~0.6)	0.5(0.4~0.6)	0.0(−0.1~0.1)	0.85
T1slope	29.3(26.9~32.3)	23.3(20.3~26.3)	6.0(1.7~10.3)	0.01 *
Lumbar lordosis	32.7(21.2~44.2)	36.2(27.8~44.6)	−3.5(−20.2~13.2)	0.64
sacral slope	36.0(32.6~43.8)	39.1(31.0~47.2)	−3.0(−7.7~1.6)	0.17
pelvic tilt	19.0(12.4~28.0)	20.1(-2.7~42.9)	−1.1(−16.0~13.8)	0.87
pelvic incidence	55.0(48.3~68.3)	55.6(35.4~75.7)	−0.5(−13.4~12.3)	0.93

CI, confidence interval; ROM, range of motion * Statistical significance ($p < 0.05$)

The difference between the two groups was statistically significant for the C0-7 angle ($p = 0.02$) and T1 slope ($p < 0.01$) before and after the surgery, respectively. The lordosis of cervical spine reduction compared to before the surgery was statistically significantly different between the two groups (showing a greater decrease in the non-lordosis group). No statistically significant difference was observed in the other cervical and spinopelvic parameters (Table 4).

3.1.3. Comparison of Pre-Operative vs. Post-Operative Radiological Parameters in the Lordosis Group

Comparison of pre-operative vs. post-operative radiological parameters in the lordosis group is represented on Figures 6 and 7.

In the lordosis group, the significant changes in the parameters before and after surgery were observed in C0-2 angle ($p < 0.01$), C0-7 angle ($p = 0.03$), C0-2ROM ($p = 0.02$), C2-7ROM ($p < 0.01$), C0-7ROM ($p < 0.01$), C0-2/C0-7 ($p < 0.01$), and C2-7/C0-7 ($p < 0.01$) (Table 5). After the surgery, the angle of c-spine (C0-7 angle) was maintained as lordotic, and C0-7ROM and C2-7ROM decreased, while C0-2 ROM and C0-2/C0-7ROM increased. These results suggest that the ROM of the cervical motion, which was reduced in the lordosis group after the surgery, was compensated through the upper cervical motion (C0-2). No statistically significant difference was observed in the other cervical and spinopelvic parameters (Table 5).

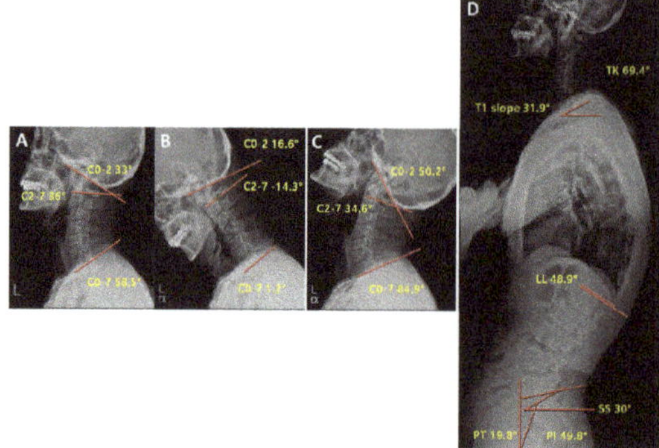

Figure 6. Pre-operative radiological parameters in lordosis groups. Cervical spine lateral X-ray of neutral (**A**), flexion (**B**), and extension (**C**), and whole-spine lateral X-ray (**D**). C0-2; C0-2 Cobb angle, the occipito-cervical angle, which is the intersection angle between the McGregor line and the line parallel to the C2 lower endplate. C2-7; C2-7 Cobb angle, the intersection angle between the line perpendicular to the line parallel to the C2 lower endplate and the line perpendicular to the line parallel to the C7 lower endplate. C0-7; C0-7 Cobb angle, the intersection angle between the McGregor line and the line parallel to the C7 lower endplate. TK; thoracic kyphosis, intersection angle between the line perpendicular to the line parallel to the T1 upper endplate and the line perpendicular to the line parallel to the T12 lower endplate. LL; lumbar lordosis, intersection angle between the line perpendicular to the line parallel to the L1 upper endplate and the line perpendicular to the line parallel to the L5 lower endplate. SS; sacral slope, the angle formed by a line drawn along the endplate of the sacrum and a horizontal reference line. PT; pelvic tilt, the angle formed by a line drawn from the midpoint of the sacral endplate to the center of the bicoxofemoral axis and a vertical and a vertical plumb line. PI; pelvic incidence, the angle formed by two vectors: the line joining the bicoxo-femoral axis to the center of the sacral end plate and the line perpendicular to the sacral endplate.

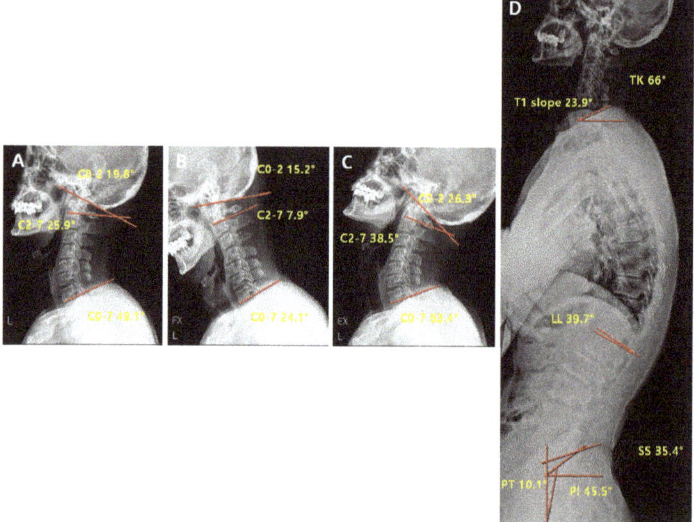

Figure 7. Post-operative radiological parameter in lordosis groups. Cervical spine lateral X-ray of neutral (**A**), flexion (**B**), extension (**C**) and whole-spine lateral X-ray (**D**).

Table 5. Comparison of pre-operative vs. post-operative radiological parameters of lordosis groups.

Parameter	Lordosis			
	Pre-op	Post-op	Between-Group Difference Mean (Degrees) (95% CI)	*p*-Value
	Mean (Degrees)	Mean (Degrees)		
C2-7angle	13.2(10.9~15.9)	11.3(8.4~14.4)	2.0(0.6~4.6)	0.13
C0-2angle	24.2(21.2~26.8)	27.2(24.2~30.5)	−3.4(−5.4~−1.3)	<0.01 *
C0-7angle	36.8(33.3~40.4)	39.7(36.1~43.7)	−3.1(−5.9~−0.3)	0.03 *
C0-2ROM	17.8(15.3~20.1)	21.7(18.6~24.7)	−4.0(−7.2~−0.8)	0.02 *
C2-7ROM	34.3(30.7~37.6)	20.9(18.1~23.9)	13.1(9.8~16.4)	<0.01 *
C0-7ROM	52.4(47.7~56.4)	42.8(39.2~46.6)	9.2(4.7~13.7)	<0.01 *
C0-2/C0-7	0.3(0.3~0.4)	0.5(0.4~0.6)	−0.2(−0.2~−0.1)	<0.01 *
C2-7/C0-7	0.7(0.6~0.7)	0.5(0.4~0.6)	0.2(0.1~0.2)	<0.01 *
T1slope	29.7(27.2~32.4)	29.3(26.9~32.3)	0.2(−2.1~2.4)	0.89
Lumbar lordosis	37.6(25.9~49.3)	32.7(21.2~44.2)	−1.1(−6.4~4.3)	0.57
sacral slope	34.0(29.6~38.5)	36.0(32.6~43.8)	−0.7(−10.8~9.5)	0.85
pelvic tilt	15.5(12.1~18.8)	19.0(12.4~28.0)	1.2(−10.1~12.4)	0.77
pelvic incidence	47.4(40.8~53.9)	55.0(48.3~68.3)	0.8(−14.0~15.6)	0.88

CI, confidence interval; ROM, range of motion. * Statistical significance ($p < 0.05$).

3.1.4. Comparison of Pre-Operative vs. Post-Operative Radiological Parameters in the Non- Lordosis Group

In the non-lordosis group, the significant changes in the parameters before and after surgery were observed in C2-7ROM ($p < 0.01$), C0-7ROM ($p < 0.01$), and T1 slope (Table 6). Furthermore, as in the lordosis group, C2-7ROM and C0-7ROM decreased after the surgery, while C0-2ROM did not increase. These results suggest that the additional decrease in ROM after the surgery could not be compensated by the upper cervical spine (C0-2) because C0-2ROM was already heavily used before the surgery. In addition, change in T1 slope (decrease) was observed after the surgery, which means more kyphotic curvature (loss of lordosis) after surgery. No statistically significant difference was observed in the other cervical and spinopelvic parameters (Table 6).

Table 6. Comparison of pre-operative vs. post-operative radiological parameters of non-lordosis groups.

Parameter	Non-Lordosis			
	Pre-op	Post-op	Between-Group Difference MEAN (Degrees) (95% CI)	*p*-Value
	Mean (Degrees)	Mean (Degrees)		
C2-7angle	1.1(−2.6~4.9)	−0.1(−4.9~4.8)	1.2(2.6~5.0)	0.53
C0-2angle	31.9(27.9~35.9)	33.7(29.8~37.7)	−1.9(−5.1~1.4)	0.26
C0-7angle	33.4(29.6~37.3)	32.9(29.5~36.3)	0.6(−3.6~4.7)	0.78
C0-2ROM	23.1(16.2~30.1)	24.8(20.9~28.7)	−1.7(−9.5~6.2)	0.67
C2-7ROM	29.7(24.9~34.5)	19.8(14.6~24.9)	9.9(4.5~15.4)	<0.01 *
C0-7ROM	50.1(44.9~55.3)	40.9(35.2~46.7)	9.2(2.6~15.8)	<0.01 *
C0-2/C0-7	0.5(0.3~0.7)	0.6(0.5~0.7)	−0.1(−0.3~0.1)	0.14
C2-7/C0-7	0.6(0.5~0.7)	0.5(0.4~0.6)	0.1(−0.02~0.2)	0.1
T1slope	26.9(23.5~30.3)	23.3(20.3~26.3)	3.6(0.1~7.0)	0.04 *
Lumbar lordosis	26.4(12.3~39.9)	36.2(27.8~44.6)	2.4(−4.0~8.9)	0.35
sacral slope	30.9(21.7~40.1)	39.1(31.0~47.2)	−0.7(−4.3~3.0)	0.64
pelvic tilt	16.5(9.6~23.4)	20.1(−2.7~42.9)	−0.9(−13.6~11.8)	0.85
pelvic incidence	41.6 (23.8~50.5)	55.6(35.4~75.7)	−4.1(−18.9~10.7)	0.49

CI, confidence interval; ROM, range of motion * Statistical significance ($p < 0.05$).

3.1.5. Comparison of Pre-Operative vs. Post-Operative Radiological Parameters in the Reducible Non-Lordosis Group

In the reducible non-lordosis group, the significant changes in parameters before and after surgery were observed in C2-7 ROM ($p < 0.01$), C0-7 ROM ($p < 0.01$), and pelvic tilt ($p = 0.02$) (Table 7). However, no statistically significant difference was observed in the C0-2 ROM. These results suggest that the overall ROM decreases after the surgery, but not to an extent that requires compensation at the C0-2 site for maintaining the horizontal gaze or the ROM of cervical spine. No statistically significant differences were observed in the other cervical and spinopelvic parameters (Table 7). Comparison of pre-operative vs. post-operative radiological parameters in the reducible non-lordosis group is represented on Figures 8 and 9

Table 7. Comparison of pre-operative vs. post-operative radiological parameters of the reducible non-lordosis group.

Parameter	Reducible Non-Lordosis			
	Pre-op	**Post-op**	**Between-Group Difference Mean (Degrees) (95% CI)**	*p*-**Value**
	Mean (Degrees)	**Mean (Degrees)**		
C2-7angle	2.0(−2.3~6.2)	2.5(−2.9~7.9)	−0.5(−4.8~3.8)	0.81
C0-2angle	29.9(25.5~34.4)	31.5(27.6~35.5)	−1.6(−5.6~2.4)	0.42
C0-7angle	32.2(27.7~36.6)	33.3(29.2~37.4)	−1.1(−6.1~3.8)	0.64
C0-2ROM	23.7(−14.8~32.5)	23.6(19.4~27.9)	0.03(−10.0~10.0)	0.99
C2-7ROM	30.3(24.2~36.4)	18.7(12.3~25.2)	11.6(5.2~18.0)	<0.01 *
C0-7ROM	51.5(45.1~57.9)	37.9(32.0~43.8)	13.6(6.7~20.5)	<0.01 *
C0-2/C0-7	0.5(0.2~0.7)	0.6(0.5~0.8)	−0.2(−0.4~0.1)	0.20
C2-7/C0-7	0.6(0.5~0.7)	0.5(0.3~0.7)	−0.2(−0.4~0.1)	0.27
T1slope	26.3(22.1~30.5)	25.6(22.6~28.6)	0.7(2.5~3.9)	0.67
Lumbar lordosis	37.0(35.0~39.0)	32.4(25.4~39.4)	4.7(76.0~85.3)	0.60
sacral slope	40.9(36.5~47.3)	40.8(38.7~42.9)	0.1(−39.3~39.5)	0.98
pelvic tilt	13.3(8.1~18.5)	15.3(10.1~20.5)	−2.0(−2.6~-1.3)	0.02 *
pelvic incidence	54.8(44.5~65.1)	50.9(49.9~51.9)	3.9(−79.4~87.1)	0.66

CI, confidence interval; ROM, range of motion * Statistical significance ($p < 0.05$).

3.1.6. Comparison of Pre-Operative vs. Post-Operative Radiological Parameters in the Non-Reducible Non-Lordosis Group

In the non-reducible non-lordosis group, no significant change in the c-spine parameters was observed before and after the surgery, and only the T1 slope decreased significantly (Table 8). These results suggest that no compensation for alignment or ROM before the surgery occurred at any sites after the surgery, and that the kyphosis of cervical spine was maintained after the surgery as the T1 slope related to the lordosis of cervical spine continues to significantly decrease. Thus, this group may not be a good indication for laminoplasty in comparison to the other groups because the application of laminoplasty does not change the kyphosis, hardly effective, in terms of all the credits from surgery including the decompression of neural tubes through posterior migration of spinal cord as well as the recovery of blood flow along the spinal cord. No statistically significant differences were observed in the other cervical and spinopelvic parameters (Table 8). Comparison of Pre-Operative vs. Post-Operative Radiological Parameters in the Non-Reducible Non-Lordosis Group (Figures 10 and 11).

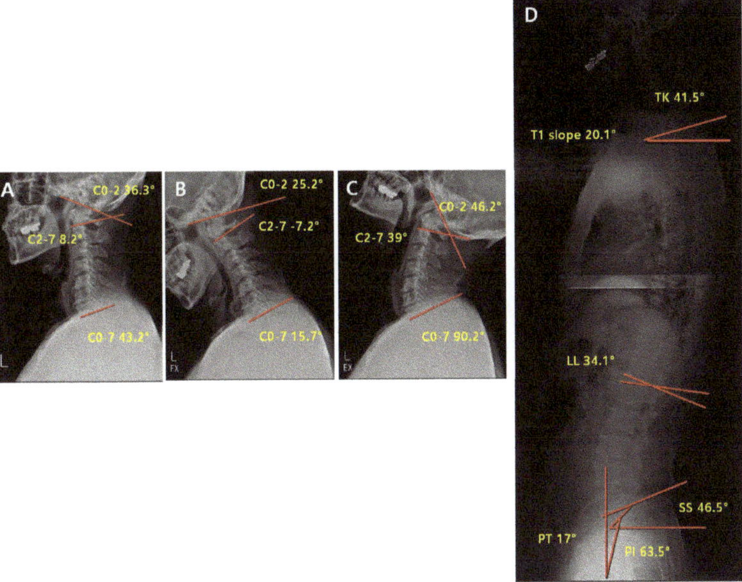

Figure 8. Pre-operative radiological parameters in the reducible non-lordosis group. Cervical spine lateral X-ray of neutral (**A**), flexion (**B**), extension (**C**) and whole-spine lateral X-ray (**D**).

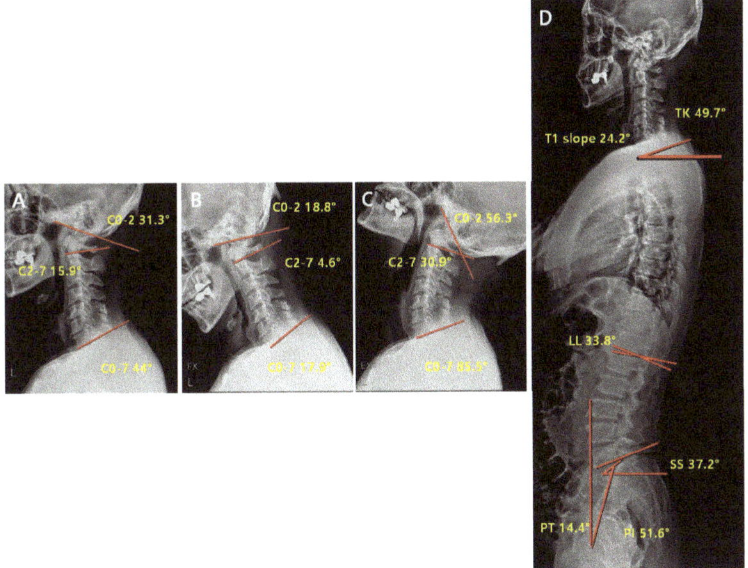

Figure 9. Post-operative radiological parameters in reducible non-lordosis group. Cervical spine lateral X-ray of neutral (**A**), flexion (**B**), extension (**C**) and whole-spine lateral X-ray (**D**).

Table 8. Comparison of pre-operative vs. post-operative radiological parameters of the non-reducible non-lordosis group.

Parameter	Nonreducible Non-Lordosis			
	Pre-op	Post-op	Between-Group	*p*-Value
	Mean (Degrees)	Mean (Degrees)	Difference Mean (Degrees) (95% CI)	
C2-7angle	−1.8(−12.5~8.8)	−9.0(−19.5~−1.6)	7.1(−1.8~16.0)	0.10
C0-2angle	38.7(29.7~47.8)	41.5(30.0~53.0)	−2.8(−10.1~4.6)	0.37
C0-7angle	37.9(28.9~47.0)	31.5(24.3~38.6)	6.5(1.5~14.4)	0.09
C0-2ROM	21.4(12.2~30.5)	28.9(16.6~41.3)	−7.6(−16.4~1.3)	0.08
C2-7ROM	27.5(20.6~34.5)	23.3(15.2~31.4)	4.2(−8.5~16.9)	0.43
C0-7ROM	45.1(36.0~54.2)	51.4(34.7~68.1)	−6.3(−19.7~7.1)	0.28
C0-2/C0-7	0.5 (0.3~0.7)	0.5(0.4~0.7)	−0.1(−0.2~0.1)	0.24
C2-7/C0-7	0.6(0.4~0.8)	0.5(0.4~0.5)	0.2(0.0~0.4)	0.09
T1slope	28.8(22.9~34.7)	15.2(9.7~20.6)	13.7(7.1~20.2)	<0.01 *
Lumbar lordosis	31.3(16.2~46.4)	28.0(18.1~37.9)	3.0(−4.1~10.1)	0.28
sacral slope	33.4(22.2~44.6)	35.0(26.9~43.1)	−0.4(−4.5~3.8)	0.79
pelvic tilt	12.8(5.0~20.6)	15.0(9.2~20.8)	−2.8(−23.4~17.7)	0.69
pelvic incidence	46.1(33.1~59.1)	41.7(20.7~62.7)	−3.2(−25.1~18.7)	0.67

CI, confidence interval; ROM, range of motion. * Statistical significance (*p* < 0.05).

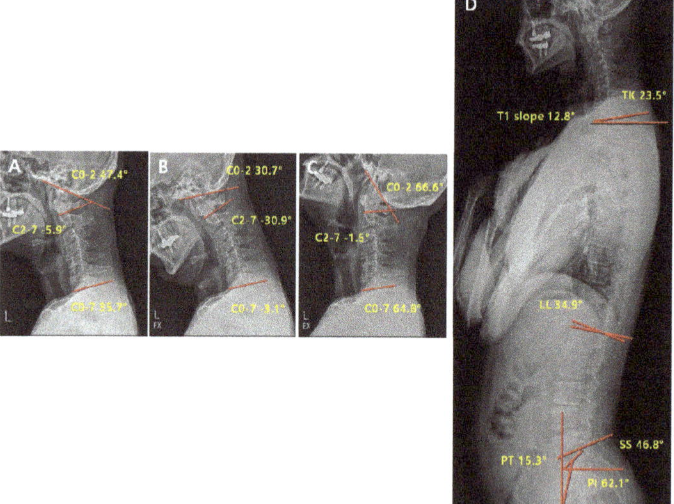

Figure 10. Pre-operative radiological parameters in the non-reducible non-lordosis group. Cervical spine lateral X-ray of neutral (**A**), flexion (**B**), extension (**C**) and whole-spine lateral X-ray (**D**).

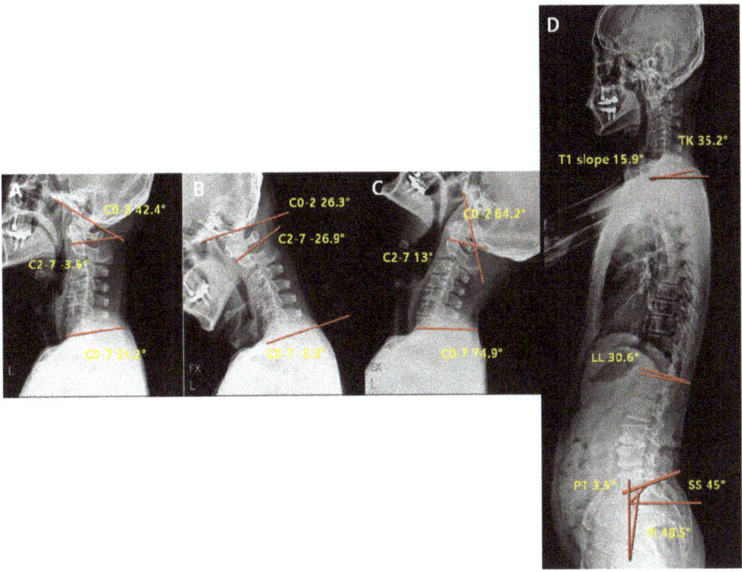

Figure 11. Post-operative radiological parameters in the non-reducible non-lordosis group. Cervical spine lateral X-ray of neutral (**A**), flexion (**B**), extension (**C**) and whole-spine lateral X-ray (**D**).

4. Discussion

Laminoplasty is an effective surgical method commonly applied in patients with myelopathy due to multiple-segment ossification of the posterior longitudinal ligament (OPLL) and spinal stenosis of cervical spine. In comparison to laminectomy and fusion surgery, laminoplasty is advantageous for DCM patients due to multiple OPLL or C-spine Herniated Nucleus Pulposus (HNP). The cervical motion can be maintained by preserving the posterior neck structure, such as the facet joint because neural tubes are extended on the laminofacet junction, where complications related to the use of instruments and direct adhesion between dura and overlying neck muscles can be reduced, and the recovery after surgery can be relatively prompt [1,4,5].

In general, it has been reported that laminoplasty is not effective for multiple-segment ossification of the posterior longitudinal ligament if kyphosis is present because indications for laminoplasty can benefit from decompression due to posterior migration of spinal cord only by maintaining the cervical lordosis [2,3,6,7,12].

However, studies have shown that kyphosis within 10° of cervical kyphosis has no different postoperative outcome from the cervical lordosis after laminoplasty [8]. Furthermore, a recent study has examined the cervical alignment of 958 normal asymptomatic individuals, finding that the ratio of the subjects with kyphotic alignment of cervical spine comprises 26.3%. The group with kyphotic alignment can be subdivided into reducible kyphosis, in which the kyphotic curvature recovers to lordotic alignment in the lateral radiograph upon extension, and non-reducible kyphosis, in which the kyphotic alignment is sustained upon extension. Although no distinctive cervical kyphosis features were observed between the two groups (reducible and nonreducible) in the resting neutral position, the feature of the motion between the two groups showed a difference in the flexion–extension ROM, depending on whether the upper cervical spine (C0–C2) or the lower cervical spine (C2–C7) ROM was compensated for the entire cervical motion. When the motion segments were divided into the upper cervical spine (C0–C2) and the lower cervical spine (C2–C7), each segment showed a distinctive ROM that served to preserve spinal functioning, especially horizontal gazing of the cervical spine. Also, this study identified that the correlation between the cervical spine and the global spine parameters (i.e.,

TK, LL, SS, PT, and PI) was not statistically significant, explaining the proper cervical spine alignment or ROM that moves the spine according to its relations with the head as the center is more necessary than focusing on the correlation between the cervical spine and the spinopelvic parameters [9].

In this regard, the authors conducted this study on the assumption that even if laminoplasty is performed in patients with cervical kyphosis, no kyphotic change is made, or the kyphotic change is minimized due to compensation of upper cervical motion (C0-2ROM) around the head center, and that the kyphosis where this upper cervical motion remains cannot be a contraindication for laminoplasty.

This study classified the curvature of the preoperative cervical spine into lordosis, reducible non-lordosis (kyphosis), and non-reducible non-lordosis (kyphosis) groups for the cases operated with cervical laminoplasty, respectively. Subsequently, this study examined the relationship between the cervical parameters and other spinopelvic parameters, and it further analyzed whether the characteristics of the curves changed after the surgery, and into which type the change would lead if it changed. Furthermore, this study analyzed whether the cervical parameter affected other spinopelvic parameters because these changes could determine whether cervical kyphosis would be a contraindication for laminoplasty in terms of radiological analysis.

The results of this study showed no difference in spinopelvic parameters between the lordotic group and the non-lordotic group before the surgery (Table 3). The results further indicated that the cervical curvature was independent of spinopelvic parameters regardless of curve characteristics of cervical spine. The results were consistent with those of other studies [8,13–15]. Moreover, the C2-7 angle before and after the surgery showed no statistically significant changes in both lordosis and non-lordosis groups (lordosis group $p = 0.13$, non-lordosis group $p = 0.53$).

There is one issue to discuss in this context. Many studies have typically measured the C2-7 angle in determining the overall ROM of the cervical spine. However, according to a recent study [9], in addition to the motion of C2-7, which is typically emphasized, the motion of C0-2 (upper cervical motion) plays an important role in the mobility of cervical spine. Thus, the authors have additionally measured C0-2 angle and ROM, C0-7 angle and ROM, as well as C2-7 angle and ROM to consider the upper cervical spine motion (C0-2) around the head center in this study.

The change before and after the surgery was examined by considering the above parameters in each group. The patients belonging to the lordosis group before the surgery showed no significant change in C2-7 angle after the surgery ($p = 0.13$), while exhibiting C0-7 angle changed to be more lordotic (between group difference mean=3.1, 95%CI: 1.3–5.4, $p=0.03$). C0-7ROM (between group difference mean= 9.2, 95%CI: 13.7 to 4.7, $p < 0.01$), and C2-7ROM (between group difference mean = 13.1, 95%CI: 16.4 to 9.8, $p < 0.01$) decreased while C0-2ROM increased (between group difference mean = 4.0, 95%CI: 0.8–7.2, $p < 0.01$). These results suggest that the ROM of cervical spine, which was reduced after the surgery, was compensated through the upper cervical spine motion (C0-2) around the head center. For this reason, the C0-2/C0-7 values also increased (between-group difference mean = −0.2, 95%CI: −0.2 to −0.1, $p < 0.01$) (Table 5).

Furthermore, according to the change in the postoperative curvature, only one of the patients belonging to the lordosis group before the surgery changed to the non-reducible non-lordosis group after the surgery, and the remaining patients maintained lordosis after surgery or its recovery upon extension even after the loss of lordosis (Figure 5).

Thus, in the patients with the lordotic curve of cervical spine before the surgery, the lordosis was mostly maintained or compensated after the surgery, and the overall ROM of the cervical spine (C2-7, C0-7 ROM) was inevitably reduced due to posterior approach surgery, while the upper cervical spine motion (C0-2 ROM, C0-2/0-7 ROM) acted in a compensatory way around the head center, resulting in maintenance of horizontal gaze and daily activities after the surgery. Accordingly, the results suggest that lordosis can be a good indication for the surgery because the effective decompression and posterior migration of the spinal cord after the surgery would recover the blood flow.

On the other hand, the patients belonging to the non-lordosis (kyphosis) group before the surgery showed a decrease in C2-7ROM and C0-7ROM after the surgery, as in the lordosis group, while exhibiting no increase in C0-2ROM unlike the lordosis group (Table 6). This suggests that there was a limit or inability to compensate for the additional decrease of ROM with the upper cervical motion (C0-2 ROM) because the C0-2ROM was already much used before the surgery. Unlike the lordosis group, the T1 slope, which indicates the lordosis of cervical spine, also tended to decrease after the surgery.

The non-lordosis (kyphosis) group was divided into reducible and non-reducible groups according to the recovery of lordosis upon extension of cervical spine to analyze the change in parameters before and after the surgery: in the reducible non-lordosis group, C2-7ROM ($p < 0.01$) and C0-7ROM ($p < 0.01$) significantly decreased, while C0-2 ROM remained unchanged. However, in the non-reducible non-lordosis group, only the T1 slope showed a significant decrease, indicating that the kyphotic state proceeded after the surgery; no changes in other parameters were observed. The change in cervical parameters, whether it was increase or decrease, was observed in all the groups other than the non-reducible non-lordosis group, while all the values including cervical spine angle and ROM were maintained after the surgery only in the non-reducible non-lordosis group. The results suggest no variability in cervical motion and angle after the surgery unlike other groups. In other words, in the non-reducible non-lordosis group, the spine move in the state C0-2, C2-7, C0-7, angle and ROM are already fixed around the head center before surgery, and the ROM of cervical spine remains unchanged without any improvement after surgery, which is not compensated by the upper cervical motion (C0-2) after the surgery.

Furthermore, the curvature changes after surgery in the non-lordosis (or kyphosis) group showed that 8 (38%) of 21 patients who belonged to the reducible non-lordosis group changed to the lordosis group, 7 (33%) remained in the reducible non-lordosis group, and 6 (29%) changed into the non-reducible non-lordosis group. On the other hand, all six patients in the non-reducible non-lordosis group remained in the same group after the surgery.

Based on the above results, non-reducible characteristics remain unchanged after the surgery in the non-reducible non-lordosis group (that is, the characteristics in which lordosis does not recover during extension), showing no variability in the cervical angle and ROM, and no compensation through the upper cervical motion (C0-2 ROM). Thus, no effect of laminoplasty (improvement effect of blood flow in the spinal cord due to the posterior migration of the spinal cord) would be observable, while the preoperative characteristics remain unchanged after the surgery. On the other hand, the reducible non-lordosis group and the lordosis group mostly show variability, and the upper cervical motion (C0-2 ROM) around the head center acts in a compensatory way. Accordingly, patients who have lordosis or reducible non-lordosis can be a good indication for laminoplasty by considering these quantitative and qualitative changes in curve, angle, and ROM of cervical spine before and after the surgery, which are independent of other spinopelvic parameters. However, selective surgery is required because some limited cases among them show characteristics that are not reducible after the surgery, and further studies are necessary to determine the selection method.

Limitations

This study has the following limitations. Firstly, this study is a retrospective study of prospectively collecting data, and despite a relatively long-term observation (36.8 months on average), 28 of 111 patients were lost during the follow-up period (25%). Furthermore, the number of the entire patient group was 83, which was not small, while the entire group was divided into subgroups, and the number of each group was relatively small to show statistical significance (non-reducible non-lordosis: 6 patients). This limitation, including a small number of patients in this study, arose because laminoplasty typically involves the lordosis of the cervical spine, and laminoplasty is rarely performed for patients with kyphosis, except in special cases. Secondly, all patients included in this study underwent midline splitting double-door laminoplasty (French door type laminoplasty). For this reason, these postoperative results could differ from those of laminoplasty applying different surgical methods (e.g., unilateral expansive open door type laminoplasty). Finally, this study has radiologically analyzed how the preoperative curve of cervical spine would change after surgery, and whether the changed curve would affect cervical parameters and other spinopelvic parameters, which could be independent of clinical results of patients before and after the surgery.

5. Conclusions

Considering radiographic changes in curve, angle, and ROM of cervical spine and its relationship with other spinopelvic parameters before and after the surgery, only patients with cervical lordosis or reducible kyphosis should be considered for laminoplasty surgery.

Author Contributions: Data curation, S.B.J.; Formal analysis, M.U.L.; Investigation, J.H.L.; Methodology, H.M.L.; Resources, J.W.K.; Software, J.S.Y.; Writing—review & editing, S.W.K. All authors have read and agreed to the published version of the manuscript.

Funding: This research received no external funding.

Conflicts of Interest: The authors declare no conflicts of interest.

References

1. Kimura, I.; Shingu, H.; Nasu, Y. Long-term follow-up of cervical spondylotic myelopathy treated by canal-expansive laminoplasty. *J. Bone Joint Surg. Br. Vol.* **1995**, *77*, 956–961. [CrossRef]
2. Suk, K.S.; Kim, K.T.; Lee, J.H.; Lee, S.H.; Lim, Y.J.; Kim, J.S. Sagittal alignment of the cervical spine after the laminoplasty. *Spine* **2007**, *32*, E656–E660. [CrossRef] [PubMed]
3. Machino, M.; Yukawa, Y.; Hida, T.; Ito, K.; Nakashima, H.; Kanbara, S.; Morita, D.; Kato, F. Cervical alignment and range of motion after laminoplasty: Radiographical data from more than 500 cases with cervical spondylotic myelopathy and a review of the literature. *Spine* **2012**, *37*, E1243–E1250. [CrossRef] [PubMed]
4. Ratliff, J.K.; Cooper, P.R. Cervical laminoplasty: A critical review. *J. Neurosurg.* **2003**, *98*, 230–238. [CrossRef] [PubMed]
5. Iwasaki, M.; Okuda, S.Y.; Miyauchi, A.; Sakaura, H.; Mukai, Y.; Yonenobu, K.; Yoshikawa, H. Surgical strategy for cervical myelopathy due to ossification of the posterior longitudinal ligament: Part 2: Advantages of anterior decompression and fusion over laminoplasty. *Spine* **2007**, *32*, 654–660. [CrossRef] [PubMed]
6. Seichi, A.; Takeshita, K.; Ohishi, I.; Kawaguchi, H.; Akune, T.; Anamizu, Y.; Kitagawa, T.; Nakamura, A.K. Long-term results of double-door laminoplasty for cervical stenotic myelopathy. *Spine* **2001**, *26*, 479–487. [CrossRef] [PubMed]
7. Sodeyama, T.; Goto, S.; Mochizuki, M.; Takahashi, J.; Moriya, H. Effect of decompression enlargement laminoplasty for posterior shifting of the spinal cord. *Spine* **1999**, *24*, 1527–1531. [CrossRef] [PubMed]
8. Kim, S.W.; Hai, D.M.; Sundaram, S.; Kim, Y.C.; Park, M.S.; Paik, S.H.; Kwak, Y.-H.; Kim, T.-H. Is cervical lordosis relevant in laminoplasty? *Spine J.* **2013**, *13*, 914–921. [CrossRef] [PubMed]
9. Kim, S.W.; Kim, T.-H.; Bok, D.H.; Jang, C.; Yang, M.H.; Lee, S.; Yoo, J.-H.; Kwak, Y.H.; Oh, J.K. Analysis of cervical spine alignment in currently asymptomatic individuals: Prevalence of kyphotic posture and its relationship with other spinopelvic parameters. *Spine J.* **2018**, *18*, 797–810. [CrossRef] [PubMed]
10. Fujiyoshi, T.; Yamazaki, M.; Kawabe, J.; Endo, T.; Furuya, T.; Koda, M.; Okawa, A.; Takahashi, K.; Konishi, H. A new concept for making decisions regarding the surgical approach for cervical ossification of the posterior longitudinal ligament: The K-line. *Spine* **2008**, *33*, E990–E993. [CrossRef] [PubMed]
11. Chiba, K.; Toyama, Y.; Watanabe, M.; Maruiwa, H.; Matsumoto, M.; Hirabayashi, K. Impact of longitudinal distance of the cervical spine on the results of expansive open-door laminoplasty. *Spine* **2000**, *25*, 2893–2898. [CrossRef] [PubMed]
12. Steinmetz, M.P.; Kager, C.D.; Benzel, E.C. Ventral correction of postsurgical cervical kyphosis. *J. Neurosurg.* **2003**, *98*, 1–7. [CrossRef] [PubMed]
13. Lee, N.-H.; Ha, J.-K.; Chung, J.-H.; Hwang, C.J.; Lee, C.S.; Cho, J.H. A retrospective study to reveal the effect of surgical correction of cervical kyphosis on thoraco-lumbo-pelvic sagittal alignment. *Eur. Spine J.* **2016**, *25*, 2286–2293. [CrossRef] [PubMed]

14. Lee, S.-H.; Son, E.-S.; Seo, E.-M.; Suk, K.-S.; Kim, K.-T. Factors determining cervical spine sagittal balance in asymptomatic adults: Correlation with spinopelvic balance and thoracic inlet alignment. *Spine J.* **2015**, *15*, 705–712. [CrossRef] [PubMed]

15. Sugrue, P.A.; McClendon, J.; Smith, T.R.; Halpin, R.J.; Nasr, F.F.; O'shaughnessy, B.A.; Koski, T.R. Redefining global spinal balance: Normative values of cranial center of mass from a prospective cohort of asymptomatic individuals. *Spine* **2013**, *38*, 484–489. [CrossRef] [PubMed]

Journal of
Clinical Medicine

Article

The Impact of Older Age on Functional Recovery and Quality of Life Outcomes after Surgical Decompression for Degenerative Cervical Myelopathy: Results from an Ambispective, Propensity-Matched Analysis from the CSM-NA and CSM-I International, Multi-Center Studies

Jamie R. F. Wilson [1,2], Jetan H. Badhiwala [1,2], Fan Jiang [1,2], Jefferson R. Wilson [1,3], Branko Kopjar [4], Alexander R. Vaccaro [5] and Michael G. Fehlings [1,2,*] on behalf of Investigators from the AOSpine North America and CSM-International Studies

[1] Division of Neurosurgery, Department of Surgery, University of Toronto, Toronto, ON M5T2S8, Canada; jamie.wilson@mail.utoronto.ca (J.R.F.W.); jetan.badhiwala@medportal.ca (J.H.B.); frank.jiang@mail.utoronto.ca (F.J.); WilsonJeff@smh.ca (J.R.W.)
[2] Division of Neurosurgery, Toronto Western Hospital, University Health Network, Toronto, ON M5T2S8, Canada
[3] Division of Neurosurgery, St. Michael's Hospital, University Health Network, Toronto, ON M5B1W8, Canada
[4] Department of Health Services, School of Public Health, University of Washington, Seattle, WA 98195, USA; branko.kopjar@nor-consult.com
[5] Department of Orthopedic Surgery, The Rothman Institute at Thomas Jefferson University, Philadelphia, PA 19107, USA; alex.vaccaro@rothmaninstitute.com
*   Correspondence: Michael.Fehlings@uhn.ca; Tel.: +1-416-603-5627

Received: 20 August 2019; Accepted: 13 October 2019; Published: 17 October 2019

Abstract: Background: The effect on functional and quality of life (QOL) outcomes of surgery in elderly degenerative cervical myelopathy (DCM) patients has not been definitively established. Objective: To evaluate the effect of older age on the functional and QOL outcomes after surgery in an international, multi-center cohort of patients with DCM. Methods: 107 patients aged over 70 years old (mean 75.6 ± 4.4 years) were enrolled in the AOSpine CSM-North America and International studies. A propensity-matched cohort of 107 patients was generated from the remaining 650 adults aged <70 years old (mean 56.3 ± 9.6 years), matched to gender, complexity of surgery, co-morbidities, and baseline functional impairment (modified Japanese Orthopedic Association scale (mJOA). Functional, disability, and QOL outcomes were compared at baseline and at two years post-operatively, along with peri-operative adverse events. Results: Both cohorts were equivalently matched. At two years, both cohorts showed significant functional improvement from the baseline but the magnitude was greater in the younger cohort (mJOA 3.8 (3.2–4.4) vs. 2.6 (2.0–3.3); $p = 0.007$). This difference between groups was also observed in the SF-36 physical component summary (PCS) and mental component summary (MCS) outcomes ($p = <0.001$, $p = 0.007$), but not present in the neck disability index (NDI) scores ($p = 0.094$). Adverse events were non-significantly higher in the elderly cohort (22.4% vs. 15%; $p = 0.161$). Conclusions: Elderly patients showed an improvement in functional and QOL outcomes after surgery for DCM, but the magnitude of improvement was less when compared to the matched younger adult cohort. An age over 70 was not associated with an increased risk of adverse events.

Keywords: degenerative cervical myelopathy; elderly; old age; outcomes; complications; mJOA; SF-36

---

1. Introduction

Degenerative cervical myelopathy (DCM) is a family of non-traumatic spinal cord injuries that contribute to spinal cord compression and the progressive onset of neurological deficits [1,2]. DCM encompasses a number of related conditions such as cervical spondylotic myelopathy (CSM), ossification of the posterior longitudinal ligament (OPLL), ossification of the ligamentum flavum, degenerative disc disease, and various congenital malformations that cause stenosis or instability leading to eventual spinal cord dysfunction [2]. It is now the commonest form of spinal cord dysfunction in adults [3], and has become an increasingly important area of focus for spine surgeons and clinician scientists in recent years [4–9]. This renewed focus has primarily been driven by a greater understanding of the benefit of timely intervention [5,7,10,11], and international clinical practice guidelines for the management of DCM were published in 2017 [7].

The aging population, juxtaposed with developments in medical technologies, has led to the "epidemiological transition"; the shift away from traditional causes of disease and mortality (infectious, nutritional deficiency, parasitic) toward chronic and degenerative diseases [3]. The incidence of DCM increases with age [3]; cervical spondylosis affects half of individuals aged 50–59 years, and nearly all individuals over the age of 70 [12]. The natural history of DCM is one of progressive neurological deficit, with the potential irreversible loss of dexterity, quadriplegia, and sphincter dysfunction [3,12,13]. DCM is set to become a major cause of chronic disability and global disease burden, if the current demographic expansion continues [14].

Decompressive surgery has been shown to not only halt neurological deterioration in DCM, but can produce a significant recovery in the neurological impairment that individuals experience [7,15]. The extent of functional and quality of life (QOL) improvement after surgery is influenced by the duration of symptoms, severity of functional impairment at presentation, the presence of co-morbidities, tobacco smoking status as well as age at presentation [8,11,16–20]. Elderly patients often present with a longer duration of symptoms, and more severe functional deficit compared to younger adults with DCM [11,12,18,19]. In addition, elderly patients have a higher prevalence of degenerative conditions, co-morbidities, an increased risk of osteoporosis, and higher peri-operative morbidity and mortality when compared to younger patients [3,12,21].

Age is a common surrogate for frailty or physiological reserve. The effect that frailty alone has on the effect of spine surgery has been studied through various means, and seems most relevant in the practice of spinal deformity where the procedures are often invasive and of long duration [22–28]. However, the significance of age on the effect of surgical outcomes for DCM is unclear. Available evidence would suggest that increasing age is associated with a worse functional outcome after surgical decompression [29–32]. However, many articles based on retrospective case series have also presented evidence demonstrating no significant differences in outcomes in terms of the modified Japanese Orthopedic Association score (mJOA), Nurick, and SF-36 scores when directly compared to standardized, younger patient cohorts [33–35]. Older patients are also more likely to undergo posterior surgery with a higher number of operated cervical levels when compared to younger adults [15,16,19].

The objectives of the current study were to (1) explore the functional and QOL impairment of adults over 70 years old with DCM compared to younger adults; (2) define how the variable of age over 70 alone can contribute to functional outcomes after DCM surgery, when accounting for other common age-related variables; and (3) define how elderly age can affect the QOL outcomes when compared to an equivalent adult cohort. We hypothesized that (1) adults over 70 (when compared to a matched cohort of younger adults with the same functional impairment) would exhibit less quality of life improvement; (2) when the effects of age-related co-morbidities and surgical factors were adjusted for, age over 70 would remain a significant risk factor for worse functional improvement after surgery; and (3) younger adults would have a more sustained improvement in QOL outcomes when compared to the elderly cohort.

2. Methods

2.1. Subjects

The AOSpine CSM-NA study recruited 278 patients with symptomatic DCM and correlating magnetic resonance imaging findings over a duration of two years from 12 North American centers [15]. The AOSpine CSM-International study recruited 479 symptomatic DCM patients from six Asian, five European, three Latin American, and two North American sites over a duration of four years [17]. Both studies were prospective observational studies, and patients were eligible for enrolment if they met the following inclusion criteria: (1) symptomatic DCM with one or more clinical signs of myelopathy; (2) imaging evidence compatible with spinal cord compression; (3) aged 18 years or older; and (4) no previous history of cervical spine procedures. Exclusion criteria included patients with malignancy or metastatic disease, active systemic infection, trauma, rheumatoid arthritis (or other inflammatory disease such as ankylosing spondylitis), symptomatic tandem lumbar stenosis, or asymptomatic DCM patients. All centers obtained approval from their respective local ethical boards prior to commencement of the study.

2.2. Baseline Characteristics

All patients enrolled in both studies had demographic data recorded prior to surgery (age, race, socioeconomic status, tobacco smoking status, body mass index (BMI), presence of co-morbidities) together with a focused myelopathy history including duration of symptoms, clinical signs, and etiology of DCM. All patients underwent surgical treatment, with the surgical approach and number of spinal levels operated left to the discretion of the treating surgeon. Peri-operative demographics were recorded including the surgical procedure of choice (anterior discectomy, corpectomy, laminoplasty, laminectomy with or without instrumented fusion). Post-operative complications up to 24 months after surgery were recorded.

2.3. Outcomes

Functional and QOL assessments were performed prior to surgery at baseline, then six, 12, and 24 months after surgery. Functional status was assessed by the mJOA scale; a standardized assessment of neurological and functional impairment that was administered by investigators [4,5,36]. Disability and QOL assessments were self-reported outcome measures in the form of the Neck Disability Index (NDI—specific to cervical degenerative pathologies) and the Short Form-36 version 2 (SF-36—a generic health-related QOL measurement). The SF-36 was further separated into the SF-36 Physical Component Summary (PCS) and Mental Component Summary (MCS) in an effort to distinguish between the patient-reported perception of physical health compared to mental and emotional well-being.

2.4. Statistical Analysis

All statistical analyses were performed using Stata 15 (Stata Corp, College Station, TX, USA) and R version 3.5.1 (R Foundation for Statistical Computing, Vienna, Austria) with an a priori specified significance level of $p = 0.05$ (two-tailed). Descriptive statistics are listed as mean and standard deviation (SD) for continuous variables and count and percentage for categorical variables.

2.4.1. Propensity Score Matching

The propensity score matching algorithm was developed and executed using the 'MatchIt' package for R statistical software by one of the coauthors (J.H.B.). Variables to be included as covariates in the generation of propensity scores were defined a priori by author consensus of clinical relevance.

Propensity scores were calculated as the probability of age $\geq$ 70 years versus age < 70 years using the logit method with the baseline mJOA score, duration of DCM symptoms, cardiac disease, diabetes, smoking (current), psychiatric disease, surgical approach, and number of operated levels as covariates (independent variables). Propensity score matching was performed in a one-to-one ratio using the 'optimal matching' technique to minimize the average absolute distance across all matched pairs. This resulted in two study groups (age < 70 vs. age ≥ 70) adjusted for the baseline covariates specified above. The baseline characteristics were compared between the study groups by the t-test for continuous variables and chi-square test for categorical variables.

2.4.2. Analysis of Outcomes

Two year outcomes for functional status (mJOA score), disability (NDI), and health-related quality of life (HRQOL; SF-36 PCS, and MCS) were compared between the study groups using an analysis of covariance (ANCOVA) adjusting for baseline score. Effect sizes for each outcome measure were summarized by β coefficients (mean difference) and associated 95% confidence intervals (CIs).

3. Results

Both young and older cohorts were matched sufficiently in terms of sex, duration of symptoms, smoking history, co-morbidities, number of levels, surgical approach, and baseline functional impairment (Table 1). The mean age of the younger cohort was 56.3 ± 9.6 when compared to 75.6 ± 4.4 in the older cohort ($p < 0.001$). The younger cohort demonstrated a significantly worse baseline SF-36 Physical Component Score (PCS, 30.7 ± 8.2 vs. 33.5 ± 8.8 ($p = 0.019$)) and Mental Component Score (MCS, 38.4 ± 12.8 vs. 43.0 ± 13.1, ($p = 0.011$)) when compared to the older cohort.

Table 1. Baseline demographics of the elderly cohort and younger cohort after propensity matching for gender, duration of symptoms, smoking status, co-morbidities, surgical factors, and baseline functional impairment.

Variable	Age < 70, N = 107	Age ≥ 70, N = 107	*p* Value
Demographics			
Age (years)—mean ± SD	56.3 ± 9.6	75.6 ± 4.4	<0.001 *
Female gender—No. (%)	68 (63.6)	68 (63.6)	1.000
Duration of symptoms—No. (%)	32.1 ± 49.7	26.9 ± 37.1	0.388
Smoker—No. (%)	6 (5.6)	10 (9.3)	0.299
Diabetes—No. (%)	14 (13.1)	15 (14.0)	0.842
Cardiovascular disease—No. (%)	70 (65.4)	74 (69.2)	0.560
Pulmonary disease—No. (%)	10 (9.3)	10 (9.3)	1.000
Psychiatric disease—No. (%)	8 (7.5)	8 (7.5)	1.000
Surgical factors			
Number of levels	3.2 ± 1.3	3.3 ± 1.3	0.591
Approach			0.889
Anterior	36 (33.6)	34 (31.8)	
Posterior	66 (61.7)	69 (64.5)	
Combined	5 (4.7)	4 (3.7)	
Functional status, disability, and QOL			
mJOA score	10.9 ± 2.8	11.0 ± 2.7	0.902
NDI	43.4 ± 20.5	39.5 ± 19.5	0.164
SF-36 PCS	30.7 ± 8.2	33.5 ± 8.8	0.019 *
SF-36 MCS	38.4 ± 12.8	43.0 ± 13.1	0.011 *

mJOA = modified Japanese Orthopedic Association scale, NDI = Neck Disability Index, SF-36 PCS = SF-36 Physical Component Score, SF-36 MCS = SF-36 Mental Component Score. * *p* value < 0.05.

Both cohorts demonstrated an improvement in functional impairment at two years, as defined by the mJOA scale ($p = 0.001$, see Table 2). Significant improvements were also seen in the NDI, PCS, and MCS scores, with the exception of the elderly MCS score ($p = 0.077$). The functional outcomes in the younger cohort were of a greater magnitude when compared to the older group (mean difference 3.8 ± 3.0 versus 2.6 ± 3.3; $p = 0.007$; see Table 3). This difference was also present in the QOL measurements (PCS and MCS, $p < 0.001$, $p = 0.007$). The change in NDI scores between the groups at two years showed no significant difference ($p = 0.094$).

Table 2. Functional, disability, and quality of life assessment scores of both cohorts at the two-year interval.

	Baseline	2 Years	MD (95% CI)	*p* Value
Age < 70 years				
mJOA	10.9 ± 2.8	14.8 ± 2.5	3.8 (3.2 to 4.4)	<0.001
NDI	43.4 ± 20.5	27.4 ± 18.6	−16.1 (−19.7 to −12.5)	<0.001
SF36-PCS	30.7 ± 8.2	38.9 ± 10.8	8.3 (6.5 to 10.2)	<0.001
SF-36 MCS	38.4 ± 12.8	46.8 ± 13.0	8.0 (5.3 to 10.8)	<0.001
Age ≥ 70 years				
mJOA	11.0 ± 2.7	13.6 ± 2.9	2.6 (2.0 to 3.3)	<0.001
NDI	39.5 ± 19.5	28.2 ± 17.8	−11.4 (−15.6 to −7.2)	<0.001
SF36-PCS	33.5 ± 8.8	36.9 ± 10.0	3.4 (1.4 to 5.5)	0.001
SF-36 MCS	43.0 ± 13.1	45.7 ± 13.7	2.6 (−0.3 to 5.5)	0.077

mJOA = modified Japanese Orthopedic Association scale, NDI = Neck Disability Index, SF-36 PCS = SF-36 Physical Component Score, SF-36 MCS = SF-36 Mental Component Score, MD = Mean Difference.

Table 3. Delta values (change in the scores from baseline) at two years for younger and older cohorts for all outcomes.

Outcome	Age < 70 Years	Age ≥ 70 Years	MD (95% CI)	*p* Value
ΔmJOA	3.8 ± 3.0	2.6 ± 3.3	1.2 (0.3 to 2.1)	0.007
ΔNDI	−16.1 ± 18.4	−11.4 ± 20.9	−4.7 (−10.2 to 0.8)	0.094
ΔPCS	8.3 ± 9.5	3.4 ± 10.6	4.9 (2.2 to 7.7)	<0.001
ΔMCS	8.0 ± 14.2	2.6 ± 14.9	5.5 (1.5 to 9.4)	0.007

mJOA = modified Japanese Orthopedic Association scale, NDI = Neck Disability Index, SF-36 PCS = SF-36 Physical Component Score, SF-36 MCS = SF-36 Mental Component Score, MD = Mean Difference.

The total number of all-cause adverse events over two years, including peri-operative complications and worsening functional impairment, was lower in the younger cohort ($n = 16$; 15%) when compared to the older cohort ($n = 24$; 22.4%; $p = 0.161$), but this was not statistically significant (see Table 4). Post-operative infection and hardware failure were equivalent between the younger and older cohorts ($p = 0.701$ and $p = 0.313$, respectively). The incidence of post-operative dysphagia was significantly higher in the older cohort when compared to the younger cohort ($n = 6$ compared to $n = 0$; $p = 0.013$). The incidence of the worsening of myelopathy symptoms was higher in the older cohort when compared to the younger cohort, which was just shy of statistical significance ($n = 17$ (15.9%) vs. $n = 8$ (7.5%); $p = 0.055$).

Table 4. List of complications of elderly and younger adult cohorts over the two-year follow up.

Complication	Age < 70, N = 107	Age ≥ 70, N = 107	*p* Value
Infection	4 (3.7)	3 (2.8)	0.701
Deficit	2 (1.9)	0 (0)	0.155
CSF leak	2 (1.9)	1 (0.9)	0.561
Deformity	0 (0)	0 (0)	1.000
Hardware	3 (2.8)	1 (0.9)	0.313
Dysphagia	0 (0)	6 (5.6)	0.013 *
Dysphonia	0 (0)	1 (0.9)	0.316
Revision	0 (0)	1 (0.9)	0.316
Hematoma	0 (0)	1 (0.9)	0.316
Adjacent segment disease	1 (0.9)	0 (0)	0.316
Vascular injury	0 (0)	0 (0)	1.000
Worsening of myelopathy	8 (7.5)	17 (15.9)	0.055
Any complication	16 (15.0)	24 (22.4)	0.161

* *p* value < 0.05.

4. Discussion

Renewed focus on functional and quality of life outcomes after surgery for DCM in adults has come from rigorous scrutiny of the results of several large-scale studies in the last decade [2,4,5,7,10,17,37–42]. The traditional doctrine of the use of surgery to 'arrest clinical progression' has been replaced with a more modern approach with decompressive surgery used as a means to provide both functional and quality of life improvement [5,7]. The assessment of quality of life metrics has become particularly important in the examination of patients with mild functional impairment (or asymptomatic cord compression), and is the focus of a number of studies [5,7,43]. Careful evaluation has discovered a number of predictors of clinical outcomes from surgery for DCM, including baseline severity of impairment, duration of symptoms, presence of systemic or psychological co-morbidities, obesity, and of course age [11,18,21]. With simple univariate analysis of outcome measures, many studies have reported worse functional and QOL outcomes with increasing age after surgery for DCM [29–32]. However, elderly patients often present with increased duration and worse severity of symptoms, with an increased incidence of co-morbidities [11,18,29]. Many elderly patients also present a high, sometimes unacceptable, surgical risk profile that makes decisions regarding surgical management less appropriate. DCM is a disease that can cause a significant amount of potentially reversible or preventable neurological disability in the elderly population, and therefore concerted efforts should be made to improve decision-making strategies for this patient group.

The results of the current study demonstrate a number of important conclusions, some of which have not been described in previous literature. Firstly, even when matched for co-morbidities, duration of symptoms, sex, smoking status and severity of functional impairment, younger adults report significantly worse effects on their baseline physical and mental quality of life scores compared to the older aged cohort. This could be explained through 2 potential mechanisms. Firstly, as the younger cohort were of working age, the psychological impact of DCM may be greater if employment security is perceived to be threatened. Secondly, the elderly patients demonstrate psychological resilience or adaptations/support that mean the symptoms of DCM have a less pronounced effect on their quality of life. This is an interesting concept, and is an important consideration for clinicians managing elderly DCM patients.

All patients showed improved functional impairment scores after 2 years (see Table 2). Taking the older cohort in isolation, this is good evidence that surgery is an effective modality to produce functional improvements in elderly DCM patients. This is an important stand-alone conclusion from this study. When the change from baseline scores at 2 years is calculated, it becomes evident that the functional improvement seen after surgery is of a greater magnitude in younger patients compared to the older cohort. This conclusion is similar to those described from previous univariate analyses in prior studies [11,18,29,32,44]. Suggested mechanisms for this discrepancy include the fact that elderly patients often present with increased duration of symptoms at diagnosis, often have difficulty accessing specialist assessment and imaging, and also have less neurological plasticity or reserve compared to younger adults [11,12,18,45,46]. The level of disability as measured by the NDI was improved significantly by surgery for all patients, but the order of magnitude was similar for both age groups ($p = 0.094$).

All patients demonstrated improved SF-36 PCS and MCS from baseline at the 2-year interval, with the exception of the MCS scores in the elderly group. Both PCS and MCS metrics echoed the functional improvement in that the younger cohort had a significantly greater degree of improvement from baseline compared to the older cohort ($p < 0.001$, $p = 0.007$). This provides good evidence that surgery leads to increased physical and mental perception of quality of life outcomes in all patients undergoing surgery for DCM. It appears that older patients' mental perception of quality of life remains stable throughout the treatment period, despite having a worse functional improvement, and less magnitude of quality of life improvement overall. This study is the first to report this difference in quality of life measures between younger and older aged cohorts and has potentially large implications on pre-surgical assessment and counselling in older patients with DCM. However, age-related effects on health perception have been found to influence the SF-36 score, and the contribution from the perception of physical effects often misses a number of important determinants of overall quality of life in older persons [47,48]. For these reasons, it has been suggested that the focus of quality of life determinants in the elderly should be weighted toward mental components rather than the physical [48]. This also raises certain questions about the efficacy of functional assessment measures for DCM in the elderly, but this was beyond the scope of the current study.

All forms of adverse events over two years were higher in the elderly cohort when compared to the younger adults, but not significantly so ($p = 0.161$, see Table 4). The risk of infection, hardware failure, CSF leak, and need for revision surgery were all equivalent between groups. The incidence of dysphagia after the anterior approach was significantly greater in the elderly cohort, which has been well described [49]. The incidence of worsening functional impairment after surgery was non-significantly higher in the elderly cohort, which is also consistent with previously described studies and may be a reflection of the disease process in DCM, rather than directly related to complications from surgery [11,40,50]. These findings suggest that, contrary to popular opinion, surgery for DCM in the elderly does not carry a significantly higher risk of adverse events when compared to younger adults that are matched for co-morbidities and complexity of surgery. This is good evidence to demonstrate that age alone does not necessarily confer an increased risk of adverse events in DCM surgery, but that the other risk factors associated with older age (increased number of levels of pathology, posterior surgery, increased co-morbidities, etc.) ostensibly play more of a role in the determinant of peri-operative risk. This concept of frailty, and its association with the assessment of surgical risk, is emerging as an important tool for surgical decision making and has been a recent focus of interest in pathologies such as spinal cord injury and adult deformity surgery [51–53]. However, the impact of frailty on the outcomes from DCM appears to be less well defined.

There are limitations of the current study. The results were from the pooled analysis of two harmonized datasets, and although the data came from prospective, multi-center sources, the original studies were not designed or powered to measure the effect of old age on the outcomes after surgery for DCM. Although both cohorts were well matched (Table 1), there could exist significant heterogeneity between the groups or hidden confounders such as drug histories that may have affected the results. The use of propensity-matching methodology helps to reduce the over-fitting seen in mixed effects or regression models, but does have obvious effects in the sample size. Therefore, some aspects of the results that show trends and not significant differences may indicate that the results are underpowered in some areas.

5. Conclusions

To consider our previous hypotheses, elderly patients, when compared to younger adults (matched to functional impairment and age-related risk factors), exhibited better SF-36 PCS and MCS prior to surgery for DCM. All patients in this study demonstrated improved functional impairment two years after surgery, but the magnitude of improvement seen was greater in the younger cohort, even when baseline functional impairment and age-related risk factors were adjusted for. Elderly patients also showed improved QOL physical and mental component scores after surgery, but the extent of the increase in the physical component was reduced when compared to younger adults. Aside from the incidence of post-operative dysphagia, older age alone was not associated with a higher incidence of adverse events.

The authors believe that the results of this study provide good evidence that surgery for DCM in the elderly is effective in terms of both functional and QOL outcomes. Perhaps most importantly, these results demonstrate that the elderly DCM age group should have different expectations with regard to the extent of functional and QOL outcomes after surgery. Patients over the age of 70 with a diagnosis of DCM are likely to require specialist considerations and should be counseled appropriately with adjusted expectations. The exact degree to which each modifiable risk factor contributes to perioperative risk, and the components that affect functional and QOL outcomes, remain to be determined and should be an important focus of further research into the effects of aging on surgery for DCM. Developing a prediction model using age (or measures of frailty) and related covariate adjustment would significantly improve the calculation of the risk profile in DCM patients undergoing assessment for decompressive surgery.

Author Contributions: Conceptualization, J.R.F.W., J.H.B., J.R.W., B.K., and M.G.F.; Data curation, J.R.F.W. and B.K.; Formal analysis, J.R.F.W., J.H.B., J.R.W., and B.K.; Investigation, J.H.B., F.J., J.R.W., A.R.V., and M.G.F.; Methodology, J.R.F.W. and J.H.B.; Resources, M.G.F.; Supervision, M.G.F.; Writing—original draft, J.R.F.W., J.H.B., and F.J.; Writing—review & editing, J.R.F.W., J.H.B., F.J., J.R.W., B.K., A.R.V., and M.G.F.

Funding: The original collection of data in the CSM-NA and CSM-I studies was supported by AOSpine. J.H.B. is supported by a Fellowship Award from the Canadian Institutes of Health Research (CIHR).

Acknowledgments: M.G.F. would like to acknowledge the support from the Gerry and Tootsie Halbert Chair in Neural Repair and Regeneration and the DeZwirek Family Foundation.

Conflicts of Interest: The authors declare no relevant conflicts of interest.

References

1. New, P.W.; Cripps, R.A.; Bonne Lee, B. Global maps of non-traumatic spinal cord injury epidemiology: Towards a living data repository. *Spinal Cord* **2014**, *52*, 97–109. [CrossRef] [PubMed]

2. Nouri, A.; Tetreault, L.; Singh, A.; Karadimas, S.K.; Fehlings, M.G. Degenerative Cervical Myelopathy: Epidemiology, Genetics, and Pathogenesis. *Spine* **2015**, *40*, E675–E693. [CrossRef] [PubMed]

3. Fehlings, M.G.; Tetreault, L.; Nater, A.; Choma, T.; Harrop, J.; Mroz, T.; Santaguida, C.; Smith, J.S. The Aging of the Global Population: The Changing Epidemiology of Disease and Spinal Disorders. *Neurosurgery* **2015**, *77* (Suppl. 4), S1–S5. [CrossRef] [PubMed]

4. Badhiwala, J.H.; Witiw, C.D.; Nassiri, F.; Akbar, M.A.; Jaja, B.; Wilson, J.R.; Fehlings, M.G. Minimum Clinically Important Difference in SF-36 Scores for Use in Degenerative Cervical Myelopathy. *Spine* **2018**. [CrossRef]

5. Badhiwala, J.H.; Witiw, C.D.; Nassiri, F.; Akbar, M.A.; Mansouri, A.; Wilson, J.R.; Fehlings, M.G. Efficacy and Safety of Surgery for Mild Degenerative Cervical Myelopathy: Results of the AOSpine North America and International Prospective Multicenter Studies. *Neurosurgery* **2018**. [CrossRef]

6. Buell, T.J.; Buchholz, A.L.; Quinn, J.C.; Shaffrey, C.I.; Smith, J.S. Importance of Sagittal Alignment of the Cervical Spine in the Management of Degenerative Cervical Myelopathy. *Neurosurg. Clin. N. Am.* **2018**, *29*, 69–82. [CrossRef]

7. Fehlings, M.G.; Tetreault, L.A.; Riew, K.D.; Middleton, J.W.; Aarabi, B.; Arnold, P.M.; Brodke, D.S.; Burns, A.S.; Carette, S.; Chen, R.; et al. A Clinical Practice Guideline for the Management of Patients with Degenerative Cervical Myelopathy: Recommendations for Patients with Mild, Moderate, and Severe Disease and Nonmyelopathic Patients with Evidence of Cord Compression. *Glob. Spine J.* **2017**, *7*, 70S–83S. [CrossRef]

8. Nouri, A.; Tetreault, L.; Dalzell, K.; Zamorano, J.J.; Fehlings, M.G. The Relationship Between Preoperative Clinical Presentation and Quantitative Magnetic Resonance Imaging Features in Patients with Degenerative Cervical Myelopathy. *Neurosurgery* **2017**, *80*, 121–128. [CrossRef]

9. Wilson, J.R.; Tetreault, L.A.; Kim, J.; Shamji, M.F.; Harrop, J.S.; Mroz, T.; Cho, S.; Fehlings, M.G. State of the Art in Degenerative Cervical Myelopathy: An Update on Current Clinical Evidence. *Neurosurgery* **2017**, *80*, S33–S45. [CrossRef]

10. Badhiwala, J.H.; Witiw, C.D.; Nassiri, F.; Jaja, B.N.R.; Akbar, M.A.; Mansouri, A.; Merali, Z.; Ibrahim, G.M.; Wilson, J.R.; Fehlings, M.G. Patient phenotypes associated with outcome following surgery for mild degenerative cervical myelopathy: A principal component regression analysis. *Spine J.* **2018**. [CrossRef]

11. Tetreault, L.; Palubiski, L.M.; Kryshtalskyj, M.; Idler, R.K.; Martin, A.R.; Ganau, M.; Wilson, J.R.; Kotter, M.; Fehlings, M.G. Significant Predictors of Outcome Following Surgery for the Treatment of Degenerative Cervical Myelopathy: A Systematic Review of the Literature. *Neurosurg. Clin. N. Am.* **2018**, *29*, 115–127.e135. [CrossRef]

12. Tetreault, L.; Goldstein, C.L.; Arnold, P.; Harrop, J.; Hilibrand, A.; Nouri, A.; Fehlings, M.G. Degenerative Cervical Myelopathy: A Spectrum of Related Disorders Affecting the Aging Spine. *Neurosurgery* **2015**, *77* (Suppl. 4), S51–S67. [CrossRef]

13. Badhiwala, J.H.; Wilson, J.R. The Natural History of Degenerative Cervical Myelopathy. *Neurosurg. Clin. N. Am.* **2018**, *29*, 21–32. [CrossRef]

14. Bank, T.W. DataBank: Population Estimates and Projections. Available online: http://databank.worldbank.org/data/reports.aspx?source=health-nutrition-and-population-statistics:-population-estimates-and-projections# (accessed on 20 August 2018).

15. Fehlings, M.G.; Wilson, J.R.; Kopjar, B.; Yoon, S.T.; Arnold, P.M.; Massicotte, E.M.; Vaccaro, A.R.; Brodke, D.S.; Shaffrey, C.I.; Smith, J.S.; et al. Efficacy and safety of surgical decompression in patients with cervical spondylotic myelopathy: Results of the AOSpine North America prospective multi-center study. *J. Bone Jt. Surg.* **2013**, *95*, 1651–1658. [CrossRef]

16. Asher, A.L.; Devin, C.J.; Kerezoudis, P.; Chotai, S.; Nian, H.; Harrell, F.E., Jr.; Sivaganesan, A.; McGirt, M.J.; Archer, K.R.; Foley, K.T.; et al. Comparison of Outcomes Following Anterior vs. Posterior Fusion Surgery for Patients with Degenerative Cervical Myelopathy: An Analysis from Quality Outcomes Database. *Neurosurgery* **2018**. [CrossRef]

17. Fehlings, M.G.; Ibrahim, A.; Tetreault, L.; Albanese, V.; Alvarado, M.; Arnold, P.; Barbagallo, G.; Bartels, R.; Bolger, C.; Defino, H.; et al. A global perspective on the outcomes of surgical decompression in patients with cervical spondylotic myelopathy: Results from the prospective multicenter AOSpine international study on 479 patients. *Spine* **2015**, *40*, 1322–1328. [CrossRef]

18. Tetreault, L.; Tan, G.; Kopjar, B.; Cote, P.; Arnold, P.; Nugaeva, N.; Barbagallo, G.; Fehlings, M.G. Clinical and Surgical Predictors of Complications Following Surgery for the Treatment of Cervical Spondylotic Myelopathy: Results from the Multicenter, Prospective AOSpine International Study of 479 Patients. *Neurosurgery* **2016**, *79*, 33–44. [CrossRef]

19. Wilson, J.; Jiang, F.; Fehlings, M. Clinical predictors of complications and outcomes in degenerative cervical myeloradiculopathy. *Indian Spine J.* **2019**, *2*, 59–67. [CrossRef]

20. Yagi, M.; Fujita, N.; Okada, E.; Tsuji, O.; Nagoshi, N.; Tsuji, T.; Asazuma, T.; Nakamura, M.; Matsumoto, M.; Watanabe, K. Impact of Frailty and Comorbidities on Surgical Outcomes and Complications in Adult Spinal Disorders. *Spine* **2018**, *43*, 1259–1267. [CrossRef]

21. Tetreault, L.; Kopjar, B.; Cote, P.; Arnold, P.; Fehlings, M.G. A Clinical Prediction Rule for Functional Outcomes in Patients Undergoing Surgery for Degenerative Cervical Myelopathy: Analysis of an International Prospective Multicenter Data Set of 757 Subjects. *J. Bone Jt. Surg.* **2015**, *97*, 2038–2046. [CrossRef]

22. Charest-Morin, R.; Street, J.; Zhang, H.; Roughead, T.; Ailon, T.; Boyd, M.; Dvorak, M.; Kwon, B.; Paquette, S.; Dea, N.; et al. Frailty and sarcopenia do not predict adverse events in an elderly population undergoing non-complex primary elective surgery for degenerative conditions of the lumbar spine. *Spine J.* **2018**, *18*, 245–254. [CrossRef]

23. Flexman, A.M.; Charest-Morin, R.; Stobart, L.; Street, J.; Ryerson, C.J. Frailty and postoperative outcomes in patients undergoing surgery for degenerative spine disease. *Spine J.* **2016**, *16*, 1315–1323. [CrossRef]

24. Flexman, A.M.; Street, J.; Charest-Morin, R. The impact of frailty and sarcopenia on patient outcomes after complex spine surgery. *Curr. Opin. Anaesthesiol.* **2019**. [CrossRef]

25. Miller, E.K.; Ailon, T.; Neuman, B.J.; Klineberg, E.O.; Mundis, G.M., Jr.; Sciubba, D.M.; Kebaish, K.M.; Lafage, V.; Scheer, J.K.; Smith, J.S.; et al. Assessment of a Novel Adult Cervical Deformity Frailty Index as a Component of Preoperative Risk Stratification. *World Neurosurg.* **2018**, *109*, e800–e806. [CrossRef] [PubMed]

26. Miller, E.K.; Lenke, L.G.; Neuman, B.J.; Sciubba, D.M.; Kebaish, K.M.; Smith, J.S.; Qiu, Y.; Dahl, B.T.; Pellise, F.; Matsuyama, Y.; et al. External Validation of the Adult Spinal Deformity (ASD) Frailty Index (ASD-FI) in the Scoli-RISK-1 Patient Database. *Spine* **2018**, *43*, 1426–1431. [CrossRef]

27. Miller, E.K.; Neuman, B.J.; Jain, A.; Daniels, A.H.; Ailon, T.; Sciubba, D.M.; Kebaish, K.M.; Lafage, V.; Scheer, J.K.; Smith, J.S.; et al. An assessment of frailty as a tool for risk stratification in adult spinal deformity surgery. *Neurosurg. Focus* **2017**, *43*, E3. [CrossRef]

28. Reid, D.B.C.; Daniels, A.H.; Ailon, T.; Miller, E.; Sciubba, D.M.; Smith, J.S.; Shaffrey, C.I.; Schwab, F.; Burton, D.; Hart, R.A.; et al. Frailty and Health-Related Quality of Life Improvement Following Adult Spinal Deformity Surgery. *World Neurosurg.* **2018**, *112*, e548–e554. [CrossRef]

29. Karpova, A.; Arun, R.; Davis, A.M.; Kulkarni, A.V.; Massicotte, E.M.; Mikulis, D.J.; Lubina, Z.I.; Fehlings, M.G. Predictors of surgical outcome in cervical spondylotic myelopathy. *Spine* **2013**, *38*, 392–400. [CrossRef]

30. Zhang, P.; Shen, Y.; Zhang, Y.Z.; Ding, W.Y.; Wang, L.F. Significance of increased signal intensity on MRI in prognosis after surgical intervention for cervical spondylotic myelopathy. *J. Clin. Neurosci.* **2011**, *18*, 1080–1083. [CrossRef]

31. Zhang, Y.Z.; Shen, Y.; Wang, L.F.; Ding, W.Y.; Xu, J.X.; He, J. Magnetic resonance T2 image signal intensity ratio and clinical manifestation predict prognosis after surgical intervention for cervical spondylotic myelopathy. *Spine* **2010**, *35*, E396–E399. [CrossRef]

32. Nakashima, H.; Tetreault, L.A.; Nagoshi, N.; Nouri, A.; Kopjar, B.; Arnold, P.M.; Bartels, R.; Defino, H.; Kale, S.; Zhou, Q.; et al. Does age affect surgical outcomes in patients with degenerative cervical myelopathy? Results from the prospective multicenter AOSpine International study on 479 patients. *J. Neurol. Neurosurg. Psychiatry* **2016**, *87*, 734–740. [CrossRef]

33. Kim, B.; Yoon, D.H.; Shin, H.C.; Kim, K.N.; Yi, S.; Shin, D.A.; Ha, Y. Surgical outcome and prognostic factors of anterior decompression and fusion for cervical compressive myelopathy due to ossification of the posterior longitudinal ligament. *Spine J.* **2015**, *15*, 875–884. [CrossRef]

34. Kim, T.H.; Ha, Y.; Shin, J.J.; Cho, Y.E.; Lee, J.H.; Cho, W.H. Signal intensity ratio on magnetic resonance imaging as a prognostic factor in patients with cervical compressive myelopathy. *Medicine* **2016**, *95*, e4649. [CrossRef]

35. Uchida, K.; Nakajima, H.; Takeura, N.; Yayama, T.; Guerrero, A.R.; Yoshida, A.; Sakamoto, T.; Honjoh, K.; Baba, H. Prognostic value of changes in spinal cord signal intensity on magnetic resonance imaging in patients with cervical compressive myelopathy. *Spine J.* **2014**, *14*, 1601–1610. [CrossRef]

36. Tetreault, L.; Wilson, J.R.; Kotter, M.R.; Nouri, A.; Cote, P.; Kopjar, B.; Arnold, P.M.; Fehlings, M.G. Predicting the minimum clinically important difference in patients undergoing surgery for the treatment of degenerative cervical myelopathy. *Neurosurg. Focus* **2016**, *40*, E14. [CrossRef]

37. Fehlings, M.G.; Barry, S.; Kopjar, B.; Yoon, S.T.; Arnold, P.; Massicotte, E.M.; Vaccaro, A.; Brodke, D.S.; Shaffrey, C.; Smith, J.S.; et al. Anterior versus posterior surgical approaches to treat cervical spondylotic myelopathy: Outcomes of the prospective multicenter AOSpine North America CSM study in 264 patients. *Spine* **2013**, *38*, 2247–2252. [CrossRef]

38. Fujimori, T.; Iwasaki, M.; Okuda, S.; Takenaka, S.; Kashii, M.; Kaito, T.; Yoshikawa, H. Long-term results of cervical myelopathy due to ossification of the posterior longitudinal ligament with an occupying ratio of 60% or more. *Spine* **2014**, *39*, 58–67. [CrossRef]

39. Ghobrial, G.M.; Harrop, J.S. Surgery vs Conservative Care for Cervical Spondylotic Myelopathy: Nonoperative Operative Management. *Neurosurgery* **2015**, *62* (Suppl. 1), 62–65. [CrossRef]

40. Tetreault, L.; Ibrahim, A.; Cote, P.; Singh, A.; Fehlings, M.G. A systematic review of clinical and surgical predictors of complications following surgery for degenerative cervical myelopathy. *J. Neurosurgery. Spine* **2016**, *24*, 77–99. [CrossRef]

41. Tetreault, L.; Nouri, A.; Singh, A.; Fawcett, M.; Nater, A.; Fehlings, M.G. An Assessment of the Key Predictors of Perioperative Complications in Patients with Cervical Spondylotic Myelopathy Undergoing Surgical Treatment: Results from a Survey of 916 AOSpine International Members. *World Neurosurg.* **2015**, *83*, 679–690. [CrossRef]

42. Tetreault, L.A.; Karpova, A.; Fehlings, M.G. Predictors of outcome in patients with degenerative cervical spondylotic myelopathy undergoing surgical treatment: Results of a systematic review. *Eur. Spine J.* **2015**, *24* (Suppl. 2), 236–251. [CrossRef]

43. St. Michael's Receives Nine CIHR Project Grants. Available online: http://www.stmichaelshospital.com/media/detail.php?source=hospital_news/2019/0129 (accessed on 5 June 2019).

44. Zhang, P.; Shen, Y.; Zhang, Y.Z.; Ding, W.Y. Prognosis significance of focal signal intensity change on MRI after anterior decompression for single-level cervical spondylotic myelopathy. *Eur. J. Orthop. Surg. Traumatol.* **2012**, *22*, 269–273. [CrossRef]

45. Vonck, C.E.; Tanenbaum, J.E.; Smith, G.A.; Benzel, E.C.; Mroz, T.E.; Steinmetz, M.P. National Trends in Demographics and Outcomes Following Cervical Fusion for Cervical Spondylotic Myelopathy. *Glob. Spine J.* **2018**, *8*, 244–253. [CrossRef]

46. Yamada, T.; Chen, C.C.; Murata, C.; Hirai, H.; Ojima, T.; Kondo, K.; Harris, J.R., III. Access disparity and health inequality of the elderly: Unmet needs and delayed healthcare. *Int. J. Environ. Res. Public Health* **2015**, *12*, 1745–1772. [CrossRef]

47. Bowling, A.; Banister, D.; Sutton, S.; Evans, O.; Windsor, J. A multidimensional model of the quality of life in older age. *Aging Ment. Health* **2002**, *6*, 355–371. [CrossRef]

48. Hickey, A.; Barker, M.; McGee, H.; O'Boyle, C. Measuring health-related quality of life in older patient populations: A review of current approaches. *PharmacoEconomics* **2005**, *23*, 971–993. [CrossRef]

49. Anderson, K.K.; Arnold, P.M. Oropharyngeal Dysphagia after anterior cervical spine surgery: A review. *Glob. Spine J.* **2013**, *3*, 273–286. [CrossRef]

50. Wilson, J.R.; Barry, S.; Fischer, D.J.; Skelly, A.C.; Arnold, P.M.; Riew, K.D.; Shaffrey, C.I.; Traynelis, V.C.; Fehlings, M.G. Frequency, timing, and predictors of neurological dysfunction in the nonmyelopathic patient with cervical spinal cord compression, canal stenosis, and/or ossification of the posterior longitudinal ligament. *Spine* **2013**, *38*, S37–S54. [CrossRef]

51. Ames, C.; Smith, J.; Md, P.; Pellise, F.; Kelly, M.; Alanay, A.; Acaroglu, E.; Perez-Grueso, F.; Kleinstuck, F.; Obeid, I.; et al. Artificial Intelligence Based Hierarchical Clustering of Patient Types and Intervention Categories in Adult Spinal Deformity Surgery: Towards a New Classification Scheme that Predicts Quality and Value. *Spine* **2019**, *44*, 915–926. [CrossRef]

52. Banaszek, D.; Inglis, T.; Marion, T.E.; Charest-Morin, R.; Moskven, E.; Rivers, C.S.; Kurban, D.; Flexman, A.; Ailon, T.; Dea, N.; et al. The Effect of Frailty on Outcome after Traumatic Spinal Cord Injury. *J. Neurotrauma* **2019**. [CrossRef]

53. Miller, E.; Lenke, L.G.; Espinoza-Rebmann, K.; Neuman, B.J.; Sciubba, D.M.; Smith, J.S.; Qiu, Y.; Dahl, B.; Matsuyama, Y.; Fehlings, M.G.; et al. Use of the adult spinal deformity (ASD) frailty index (ASD-FI) to predict major complications in the scoli-risk 1 multicenter, international patient database. *Spine J.* **2016**, *16*, S131–S132. [CrossRef]

Journal of
Clinical Medicine

Article

Degenerative Cervical Myelopathy in Higher-Aged Patients: How Do They Benefit from Surgery?

Oliver Gembruch [1,*], Ramazan Jabbarli [1], Ali Rashidi [1], Mehdi Chihi [1], Nicolai El Hindy [1,2], Axel Wetter [3], Bernd-Otto Hütter [1], Ulrich Sure [1], Philipp Dammann [1] and Neriman Özkan [1]

[1] Department of Neurosurgery, University Hospital Essen, University of Duisburg-Essen, 45122 Essen, Germany; ramazan.jabbarli@uk-essen.de (R.J.); ali.rashidi@med.ovgu.de (A.R.); Mehdi.Chihi@uk-essen.de (M.C.); elhindy.nicolai@krankenhaus-werne.de (N. EH.); otto.huetter@uk-essen.de (B.-O.H.); ulrich.sure@uk-essen.de (U.S.); philipp.dammann@uk-essen.de (P.D.); neriman.oezkan@uk-essen.de (N.Ö.)

[2] Spine-Center Werne, Katholisches Klinikum Lünen/Werne GmbH, St. Christophorus-Krankenhaus, Am See 1, 59368 Werne, Germany

[3] Institute of Diagnostic and Interventional Radiology and Neuroradiology, University Hospital Essen, University of Duisburg-Essen, 45122 Essen, Germany; axel.wetter@uk-essen.de

* Correspondence: oliver.gembruch@uk-essen.de; Tel.: +49-(0)201-723-1201; Fax: +49-(0)201-723-5909

Received: 20 October 2019; Accepted: 24 December 2019; Published: 26 December 2019

Abstract: Background: Degenerative cervical myelopathy (DCM) is the most common reason for spinal cord disease in elderly patients. This study analyzes the preoperative status and postoperative outcome of higher-aged patients in comparison to young and elderly patients in order to determine the benefit to those patients from DCM surgery. Methods: A retrospective analysis of the clinical data, radiological findings, and operative reports of 411 patients treated surgically between 2007 and 2016 suffering from DCM was performed. The preoperative and postoperative neurological functions were evaluated using the modified Japanese Orthopedic Association Score (mJOA Score), the postoperative mJOA Score improvement, the neurological recovery rate (NRR) of the mJOA Score, and the minimum clinically important difference (MCID). The Charlson Comorbidity Index (CCI) was used to evaluate the impact of comorbidities on the preoperative and postoperative mJOA Score. The comparisons were performed between the following age groups: G1: ≤50 years, G2: 51–70 years, and G3: >70 years. Results: The preoperative and postoperative mJOA Score was significantly lower in G3 than in G2 and G1 ($p < 0.0001$). However, the mean mJOA Score's improvement did not differ significantly ($p = 0.81$) between those groups six months after surgery (G1: 1.99 ± 1.04, G2: 2.01 ± 1.04, G: 2.00 ± 0.91). Furthermore, the MCID showed a significant improvement in every age-group. The CCI was evaluated for each age-group, showing a statistically significant group effect ($p < 0.0001$). Analysis of variance revealed a significant group effect on the delay (weeks) between symptom onset and surgery ($p = 0.003$). The duration of the stay at the hospital did differ significantly between the age groups ($p < 0.0001$). Conclusion: Preoperative and postoperative mJOA Scores, but not the extent of postoperative improvement, are affected by the patients' age. Therefore, patients should be considered for DCM surgery regardless of their age.

Keywords: degenerative cervical myelopathy; cervical canal stenosis; cervical spine surgery; higher-aged patients; neurological outcome; mJOA Score; MCID

---

1. Introduction

Degenerative cervical myelopathy (DCM) is a slowly ongoing degenerative disease of the cervical spine caused by progressive narrowing of the cervical canal and compression of the spinal cord. The DCM is age-dependent and is the most common degenerative disease of the cervical spine in

was calculated using the formula suggested by Hirabayashi and Satomi [NRR (%) = (postoperative mJOA Score–preoperative mJOA Score)/(full score (18)–preoperative mJOA Score) × 100] [18].

The mJOA Score improvement was also evaluated (postoperative mJOA Score–preoperative mJOA Score) to analyze the postoperative outcome. Comorbidities were analyzed using the CCI. A total of 27 patients (6.6%) did not attend the three-month follow-up examination and 95 patients (23.1%) were lost to follow-up six months after surgery.

Additionally, the MCID was evaluated after the operation. It is defined as the minimum change in a measurement that a patient would identify as being beneficial [19]. For DCM, the MCID is defined as follows: 1 point for patients with mild DCM (mJOA Score ≥ 15), 2 points for patients with moderate DCM (mJOA Score of 12–14), and 3 points for patients with severe DCM (mJOA Score < 12). A "poor" outcome was therefore defined as a postoperative change of mJOA Score less than the MICD.

2.3. Statistical Analysis

Data were analyzed using SPSS 23.0 (Statistical Package for the Social Sciences, SPSS Inc., Chicago, IL, USA). Metric data were described by mean and standard deviation and nominal data by frequency and valid percent. Data were checked for possible deviations from the assumption of normal distribution using the Shapiro-Wilk test. The mJOA Scores were assessed preoperatively, postoperatively, three months after surgery, and six months after surgery and were compared by Friedman-Tests for non-normally distributed data. The Man-Whitney-U-Test and the Kruskal-Wallis-Test were used to evaluate significant differences between the groups. The Wilcoxon signed-rank test was used to test between pairs of repeated measurement. Analysis of variance (ANOVA) was also used to detect statistical differences between the age groups. Pearson Chi^2 statistics were applied to compute two-sided asymptotic statistical significance. In addition, a contingency coefficient served as a measure for the symmetry of the association. McNemar-Test was used within the age groups to determine significant changes in MCID improvement within a period of six months after surgery.

2.4. Ethics

The study has been carried out in accordance with The Code of Ethics of the World Medical Association (The Declaration of Helsinki) and was approved by the Institutional Review Board (Medical Faculty, University of Duisburg-Essen, Registration number: 16-6270-BO).

3. Results

3.1. Symptom Presentation

First symptoms in group G1 were cervicobrachial neuralgia (48.6%), followed by sensory deficits (24.3%). These results were similar to those of group G2 (43.1% and 18.6%, respectively). However, the number of patients with ataxia was higher with increasing age, particularly in G3. Here, only 30.1% of the patients suffered from cervicobrachial neuralgia as the first symptom, but 40.6% complained about ataxia (Table 3).

Table 3. Analyzing the first presenting symptom in relation to the age groups.

First Symptom	G1	G2	G3
Cervicobrachial neuralgia	36 (48.6%)	88 (43.1%)	40 (30.1%)
Sensory deficit	18 (24.3%)	38 (18.6%)	23 (17.3%)
Paresis	5 (6.8%)	17 (8.4%)	16 (12.0%)
Ataxia	15 (20.3%)	61 (29.9%)	54 (40.6%)

3.2. Surgical Treatment

Surgical treatment included ACDF (243 patients, 59.22%; G1: 64; G2: 130; G3: 49), laminoplasty (117 patients, 28.40%: G1: 9; G2: 61; G3: 47), and decompressive laminectomy without posterior fusion (51 patients, 17.92%; G1: 1; G2: 13; G3: 37) (Table 4).

Table 4. Surgical treatment and complications according to the age groups; ACDF: Anterior cervical discectomy and fusion.

		G1	G2	G3
Surgical Treatment	ACDF	64	130	49
	Laminoplasty	9	61	47
	Laminectomy	1	13	37
Complications	Surgical	1/74 (1.4%)	5/204 (2.5%)	9/133 (6.8%)
	Non-surgical	1/74 (1.4%)	3/204 (1.5%)	8/133 (6.0%)

ACDF was chosen for patients with a ventral one or two-level narrowness caused by a spinal canal stenosis or a herniated disk. In patients with multilevel spinal canal stenosis, posterior decompression was favored, while laminoplasty was performed in patients with a predominantly dorsal multilevel narrowness.

However, according to the surgical approach, the number of operated levels and the surgical treatment showed no significant MCIDs six months after the operation (Table 5).

Table 5. The minimum clinically important difference (MCID) six months after surgery according to the surgical approach, the number of operated levels, and the surgical treatment; anterior cervical discectomy and fusion: ACDF.

		MCID Achievement 6 Months Postoperative		p-Value
		Yes	No	
Approach	ventral	181	14	$p = 0.669$
	dorsal	110	11	
Number of operated levels	monosegmental	162	11	$p = 0.521$
	bisegmental	69	7	
	multisegmental	60	7	
Surgical treatment	ACDF	178	14	$p = 0.282$
	Laminoplasty	84	6	
	Laminectomy	29	5	

3.3. Preoperative and Postoperative mJOA Score

The mean preoperative mJOA Score in G1 was 14.99 ± 2.17, the mean postoperative mJOA Score was 15.78 ± 2.22, while the mean mJOA Score three months after surgery was 16.58 ± 1.90 and the score six months after surgery was 17.25 ± 1.36.

The mean preoperative mJOA score in G2 was 14.57 ± 2.27, the mean postoperative mJOA score was 15.32 ± 2.47, while the mean mJOA score three months after surgery was 16.30 ± 1.92 and the score six months after surgery was 16.99 ± 1.39.

The mean preoperative mJOA Score in G3 was 13.57 ± 2.51, the mean postoperative mJOA Score was 14.23 ± 2.56, while mean the mJOA Score three months after surgery was 15.45 ± 2.02 and the mJOA Score six months after surgery was 16.32 ± 1.70.

The postoperative mJOA Score improved significantly ($p < 0.001$, respectively) in every age group, but higher-aged patients showed a lower postoperative mJOA Score compared to younger and elderly patients (Table 2 and Figure 2).

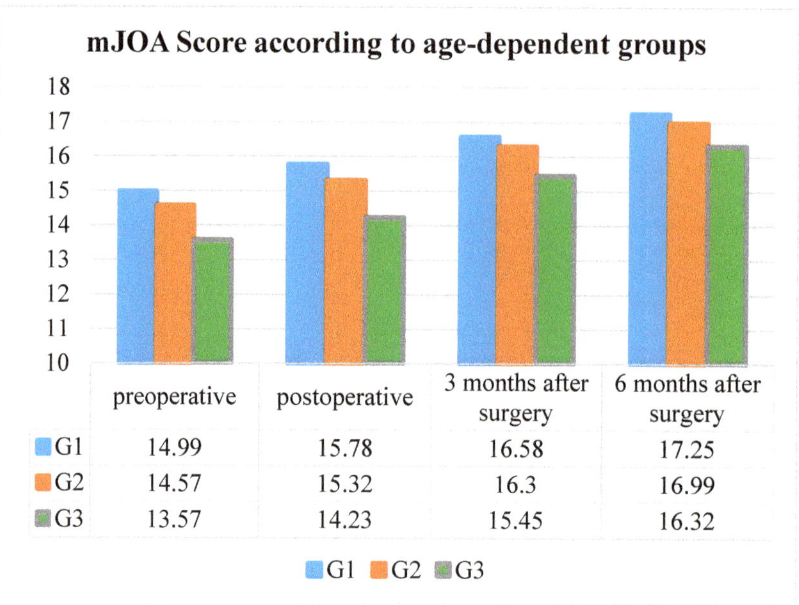

	preoperative	postoperative	3 months after surgery	6 months after surgery
G1	14.99	15.78	16.58	17.25
G2	14.57	15.32	16.3	16.99
G3	13.57	14.23	15.45	16.32

Figure 2. Modified Japanese Orthopedic Association Score (mJOA Score) according to the age-dependent groups (G1–G3).

Additionally, postoperative mJOA Scores increased significantly independently of the surgical procedure (Table 6).

Table 6. Modified Japanese Orthopedic Association Score (mJOA Score), neurological recovery rate (NRR) and the mean mJOA Score improvement according to the surgical treatment; anterior cervical discectomy and fusion: ACDF.

	Surgical Treatment	Preoperative	Postoperative	3 Months Postoperative	6 Months Postoperative	*p*-Value
mJOA Score	ACDF	14.8 ± 2.3	15.6 ± 2.4	16.6 ± 1.7	17.2 ± 1.3	<0.001
	Laminoplasty	13.9 ± 2.4	14.7 ± 2.4	15.7 ± 2.2	16.6 ± 1.6	<0.001
	Laminectomy	12.9 ± 2.3	13.8 ± 2.2	14.9 ± 1.8	15.7 ± 1.7	<0.001
mJOA Score improvement	ACDF		0.8 ± 06	1.5 ± 0.8	1.9 ± 1.0	<0.001
	Laminoplasty		0.8 ± 0.7	1.8 ± 0.9	2.3 ± 0.9	<0.001
	Laminectomy		0.8 ± 0.8	1.8 ± 1.0	2.1 ± 0.8	<0.001
NRR	ACDF		33.8 ± 33.6	61.9 ± 33.4	76.9 ± 30.2	<0.001
	Laminoplasty		25.0 ± 26.9	53.6 ± 31.4	71.6 ± 27.0	<0.001
	Laminectomy		21.4 ± 24.6	41.6 ± 23.5	54.1 ± 24.5	<0.001

3.4. Neurological Recovery Rate

The median NRR in patients according to G1 was 37.6% postoperative, 65.2% three months after surgery, and 83.4% six months after surgery in patients G1.

The median NRR in patients according to G2 was 32.9% postoperative, 60.3% three months after surgery, and 75.8% six months after surgery in patients G2.

The median NRR in patients according to G3 was 19.5% postoperative, 46.2% three months after surgery, and 61.9% six months after surgery.

NRR improved significantly in every age group at the postoperative, three and six months post-surgery follow-up examinations ($p < 0.001$, respectively). Additionally, patients belonging to G1 or G2 presented a significantly better recovery rate than patients representing G3, due to the higher preoperative mJOA Scores ($p < 0.001$) (Table 2 and Figure 3).

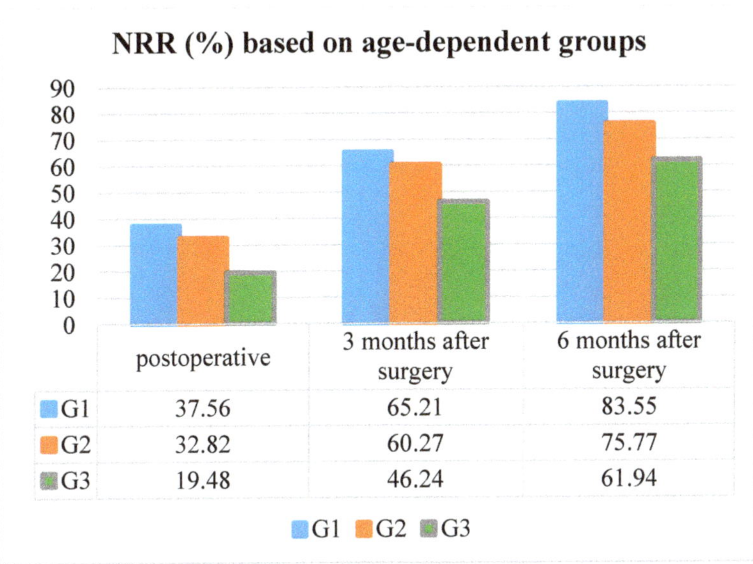

	postoperative	3 months after surgery	6 months after surgery
G1	37.56	65.21	83.55
G2	32.82	60.27	75.77
G3	19.48	46.24	61.94

Figure 3. The neurological recovery rate (NRR) based on the age-dependent groups.

Additionally, postoperative NRR increased significantly independently of the surgical procedure used (Table 6).

3.5. Mean mJOA Score Improvement

The mean mJOA Score improvement in G1 was 0.76 ± 0.79 postoperative, 1.58 ± 0.90 three months after surgery, and 1.95 ± 1.04 six months after surgery.

The mean mJOA Score improvement in G2 was 0.74 ± 0.97 postoperative, 1.60 ± 0.89 three months after surgery, and 2.01 ± 1.04 six months after surgery.

The mean mJOA Score improvement in G3 was 0.66 ± 1.02, 1.58 ± 1.13 three months after surgery, and 2.00 ± 0.91 six months after surgery.

The mean mJOA Score improvement was significantly better in every age group at the postoperative, three and six months post-surgery follow-up examinations ($p < 0.001$, respectively). However, the mean mJOA Score improvement did not differ significantly between the age groups ($p = 0.81$) (Table 2 and Figure 4).

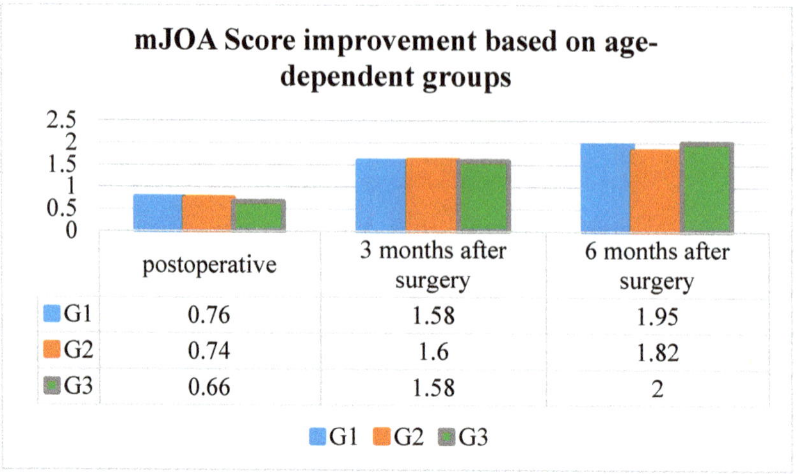

Figure 4. mJOA Score improvement based on the age-dependent groups.

Additionally, postoperative mJOA Score improvement increased significantly independently of the surgical procedure used (Table 6).

3.6. Minimum Clinically Important Difference

In our study, we were able to show favorable MCID immediately postoperatively in 52.7% of the patients of G1, in 49.0% of the patients of G2, and in 34.6% of the patients of G3. Furthermore, 98.4% (G1), 93.3% (G2), and 85.7% (G3) of the patients achieved favorable MCID six months after surgery. MCID was significantly within the different age groups. Unfortunately, the MCID achievements were significantly reduced in elderly patients compared to younger patients (Table 2).

Furthermore, the MCID six months postoperative revealed no significant differences regarding the different surgical approaches (ventral vs. dorsal), the number of operated levels (monosegmental, bisegemental, and multisegmental), and the different surgical treatment (ACDF, laminoplasty and laminectomy) (Table 5).

3.7. Comorbidities and the Charlson Comorbidity Index

Comorbidities of the patients were collected and grouped as 1. cardiovascular (e.g., arterial hypertension, coronary heart disease myocardial infarction), 2. pulmonal (e.g., chronic obstructive lung disease, pneumonia, asthma bronchiale), 3. neurological (e.g., transitory ischemic attack, stroke, polyneuropathia), 4. oncological (e.g., lung cancer, breast cancer or prostate cancer), 5. endocrine disease (e.g., thyreoditis, diabetes mellitus), and 6. surgical (abdominal, cardial, pulmonal or orthopedic).

The CCI was evaluated for each age group, showing a statistically significant group effect ($p < 0.001$) according to ANOVA. Additionally, CCI showed a significant association with the preoperative and postoperative mJOA Score ($p < 0.001$) (Table 1).

3.8. Duration of Myelopathic Symptoms Prior to Surgery

In our study, the duration lasting from the first symptom presentation until DCM surgery was much longer in G3 (45.9 ± 66.0 weeks; range: 1–350) than in G1 (22.2 ± 22.08 weeks; range: 1–122) and G2 (33.3 ± 43.6 weeks; range: 1–300). ANOVA revealed a significant group effect on the delay (weeks) between symptom onset and surgery ($p = 0.003$) (Table 1).

3.9. Duration of Hospitalization

Stay at the hospital in G1 was around 8.6 ± 3.8 days, 9.6 ± 4.0 days in G2, and 10.5 ± 4.7 days in G3. The duration of the stay at the hospital did differ significantly ($p < 0.001$) between the age-groups according to the ANOVA results (Table 1).

3.10. Surgical and Non-Surgical Complications

Complications were analyzed according to the age groups. Surgical and non-surgical complications were evaluated. Surgical complications were defined as 1. postoperative bleeding, 2. poor wound healing, 3. cerebrospinal fluid leakage, and 4. acute myelon compression. Non-surgical complications were defined as 1. pneumonia, 2. heart attack, and 3. stroke.

Non-surgical complications were highest in G3 with 6.0%, whereas those complications were similar within G1 and G2 (1.4% vs. 1.5%).

Surgical complications were reported to increase with age, affecting 1.4% (G1), 2.5% (G2), and 6.8% (G3) of the patients (Table 4).

4. Discussion

The demographic changes of the population especially in western countries [3] have led to an increase of age-dependent diseases. Therefore, understanding the treatment of elderly patients has become more relevant. At present, there are no guidelines for management of degenerative spine diseases in elderly individuals. Moreover, only a few articles have addressed this topic so far, returning partially discrepant results in functional outcomes [10–15].

Tetreault et al. showed in their systematic review that patients with a more severe DCM expressed by a lower mJOA Score and patients with a longer duration of the symptoms are more likely to have a worse surgical result. They concluded that both severe and chronic compression of the spinal cord may lead to irreversible damage due to demyelination and necrosis of the grey matter. They were also able to show that age is a potential predictor when analyzing the postoperative outcome [20]. Holly et al. also analyzed age, the duration of symptoms, and preoperative neurological function as predictors of the neurological outcome and they could show similar results in their review [21]. In our study, the time between the first symptom presentation and DCM surgery was much longer in G3 than in G1 or G2. This effect might be caused by the fact that patients in G3 suffer from more comorbidities than younger patients and that an age-dependent decrease in the daily condition might be seen as normal or might not be recognized early in older patients and therefore, the time until diagnosis is prolonged. The reduced physical condition is also expressed by the prolonged hospitalization in G3 compared to G1 and G2 and the significantly lower CCI. The CCI also showed a strong correlation with the preoperative and postoperative mJOA Scores. Additionally, the first symptom of G1 patients was cervicobrachial neuralgia, whereas the first symptom of G3 patients was ataxia followed by cervicobrachial neuralgia. This could highlight the difficulties of distinguishing DCM from other age-related diseases.

The evaluation of the preoperative mJOA Score showed significantly lower scores in G3 compared to the preoperative mJOA scores of G1 and G2. The postoperative mJOA Score and the mJOA Score three and six months after surgery showed similar results.

In our study, the significantly lower preoperative and postoperative mJOA Scores of G3 patients as compared to G2 and G1 patients was strongly associated with the lower CCI of those patients. Mean preoperative and postoperative mJOA Scores may be lower in G2 and especially in G3 due to physical weakness caused by age and known comorbidities such as cerebral vascular disorders, hip and knee osteoarthritis, entrapment of peripheral neuropathy (carpal or cubital tunnel syndrome), diabetic neuropathy, benign prostatic hypertrophy, or urinary stress incontinence [22]. Machino et al. [23] also concluded that the preoperative JOA Score might be influenced by those comorbidities.

Nagashima et al. evaluated the neurological outcomes of 37 patients over 80 years of age and compared them with that of a younger population. The NRR was lower in the elderly population, but JOA Scoring improved in a way that life style was positively influenced [10].

In the present study, there was a significant improvement according to the NNR after surgery in all age groups. Interestingly, NRR was significantly lower in G3 compared to G1 and G2 despite similar mean mJOA Score improvements.

Nevertheless, there are limitations concerning the validity of the NRR despite its popularity. The results of the NRR are strongly influenced by the preoperative mJOA Scores. For example, if patients have a low preoperative mJOA Score, then NRR is lower than in patients with a higher preoperative mJOA Score even though the mean mJOA Score improvements are the same [23].

Due to the limitations of the NRR evaluating the neurological outcome in patients with a lower preoperative mJOA Score, the mean mJOA Score improvement or the MCID might be more valuable for comparing the neurological improvements between those patients.

In our study, the mean mJOA Score improvements were similar in every age-dependent group (Table 3), although the preoperative mJOA Score was significantly lower in G3 than in G1 or G2.

The results of the mean mJOA Score improvement are in line with the findings of Machino et al. [23]. They analyzed 520 patients with CSM treated by laminoplasty and divided their patients into nonelderly (<65 years), young-old (65–75 years) and old-old (>75 years). Elderly group patients showed significantly lower recovery rates of JOA Scores compared with the nonelderly group, but mean JOA Score improvements showed no difference among these groups. Preoperative JOA Scores were also significantly lowered similarly to our patients.

Madhavan et al. also performed a meta-analysis of old DCM patients evaluating the postoperative outcome and the operative risks. They found, like in our evaluation, a significantly lower preoperative JOA Score and a lower postoperative JOA Score associated with a lower NRR. But postoperative and long-term improvements in old patients have been remarkable in terms of improvements in mobility and independence, leading to reduced nursing care being required. The incidence of postoperative complications did not show a significant difference [24].

Additionally, MCID was favorable in the majority of elderly patients (85.7%) six months after surgery. This means that those patients showed acceptable clinical improvement after surgery despite their age.

In summary, mean mJOA Score improvements did not differ significantly among the age-dependent groups, but clinical improvement after surgery according to the mJOA Score was much better in old patients compared to younger patients. This improvement led to an improvement in mobility and independence, hence requiring reduced nursing care. This was shown by Yoshida et al., who were able to show in a study with 76 patients older than 75 years of age that the nursing care requirements based on JOA Score and functional independence measure scoring was reduced [11]. Furthermore, a different surgical approach, number of operated levels, and surgical treatment revealed no significant MCID differences six months post-surgery.

Limitations

The present study has several limitations. First, this is a retrospective, non-randomized study with the associated inherent bias. The analyzed data were collected from documented electronic records, operative reports, radiological data, and reports of the patients. Secondly, the mJOA Scoring system might be influenced in the elderly group by several comorbidities such as hip and knee osteoarthritis, cerebrovascular diseases, diabetic neuropathy, or prostate hypertrophy. Additionally, the NRR system also has some limitations. Lower preoperative scores indicate lower NRR, although the mean mJOA Score improvement was the same. Furthermore, the follow-up period of six months is relatively short, caused by the retrospective nature of the study and the resulting losses in follow-up. The evaluation of the postoperative outcome might be too early directly after surgery, but we were still able to show an improvement of the neurological status over that short period.

However, future prospective studies with a longer follow-up are needed to evaluate the neurological long-term outcome in elderly patients. Nevertheless, we could show in a large population of elderly patients that surgery is still useful due to clinical improvements of the symptoms and the resulting lower need for daily care.

5. Conclusions

The preoperative and postoperative mJOA Scores are significant lower in older patients compared to younger individuals, but the mean mJOA Score improvement is similar. The lower mJOA Score in those patients correlates with a lower CCI. With respect to the postoperative mJOA Score improvement and MCID, older patients still benefit from surgery. Therefore, surgical treatment of DCM is a valuable option for those patients.

Author Contributions: Conceptualization: O.G., A.R. and N.Ö.; Data curation: O.G. and A.R.; methodology: O.G., A.R. and N.Ö.; Formal analysis: O.G., M.C. and B.-O.H. Visualization: O.G., R.J., N. EH., A.W., U.S., P.D. and N.Ö.; Writing—original draft: O.G., A.R., R.J., B.-O.H. and N.Ö.; revising it critically for important intellectual content: M.C., N. EH., A.W., U.S. and P.D.; Final approval of the version to be published: O.G., R.J., A.R., M.C., N. EH., A.W., B.-O.H., U.S., P.D. and N.Ö. All authors have read and agreed to the published version of the manuscript.

Funding: An IFORES grant (D/107-40960) to Oliver Gembruch from the University of Duisburg-Essen supported the research. The funders had no role in study design, data collection and analysis, decision to publish, or preparation of the manuscript.

Acknowledgments: We acknowledge support by the Open Access Publication Fund of the University of Duisburg-Essen.

Conflicts of Interest: The authors declare no conflict of interest.

References

1. Lawrence, J. Disc degeneration. Its frequency and relationship to symptoms. *Ann. Rheum. Dis.* **1969**, *28*, 121–138. [CrossRef]
2. United Nations, Department of Economic and Social Affairs, Population Division. World Population Ageing 2015 (ST/ESA/SER.A/390). Available online: https://www.un.org/en/development/desa/population/publications/pdf/ageing/WPA2015_Report.pdf (accessed on 28 April 2015).
3. Bevölkerung Deutschlands bis 2060. Available online: https://www.destatis.de/DE/Themen/Gesellschaft-Umwelt/Bevoelkerung/Bevoelkerungsvorausberechnung/Publikationen/Downloads-Vorausberechnung/bevoelkerung-deutschland-2060-presse-5124204159004.pdf?__blob=publicationFile (accessed on 28 April 2015).
4. Chung, S.S.; Lee, C.S.; Chung, K.H. Factors affecting the surgical results of expansive laminoplasty for cervical spondylotic myelopathy. *Int. Orthop.* **2002**, *26*, 334–338. [CrossRef]
5. Suda, K.; Abumi, K.; Ito, M.; Shono, Y.; Kaneda, K.; Fujiya, M. Local kyphosis reduces surgical outcomes of expansive open-door laminoplasty for cervical spondylotic myelopathy. *Spine (Phila Pa 1976)* **2003**, *28*, 1258–1262. [CrossRef]
6. Fehlings, M.G.; Tetreault, L.A.; Riew, K.D.; Middleton, J.W.; Aarabi, B.; Arnold, P.M.; Brodke, D.S.; Burns, A.S.; Carette, S.; Chen, R.; et al. A Clinical Practice Guideline for the Management of Patients With Degenerative Cervical Myelopathy: Recommendations for Patients With Mild, Moderate, and Severe Disease and Nonmyelopathic Patients With Evidence of Cord Compression. *Glob. Spine J.* **2017**, *7*, 70S–83S. [CrossRef]
7. Fehlings, M.G.; Tetreault, L.A.; Riew, K.D.; Middleton, J.W.; Wang, J.C. A Clinical Practice Guideline for the Management of Degenerative Cervical Myelopathy: Introduction, Rationale, and Scope. *Glob. Spine J.* **2017**, *7*, 21S–27S. [CrossRef]
8. Fehlings, M.G.; Jha, N.K.; Hewson, S.M.; Massicotte, E.M.; Kopjar, B.; Kalsi-Ryan, S. Is surgery for cervical spondylotic myelopathy cost-effective? A cost-utility analysis based on data from the AOSpine North America prospective CSM study. *J. Neurosurg. Spine* **2012**, *17*, 89–93. [CrossRef]

9. Jalai, C.M.; Worley, N.; Marascalchi, B.J.; Challier, V.; Vira, S.; Yang, S.; Boniello, A.J.; Bendo, J.A.; Lafage, V.; Passias, P.G. The Impact of Advanced Age on Peri-Operative Outcomes in the Surgical Treatment of Cervical Spondylotic Myelopathy: A Nationwide Study Between 2001 and 2010. *Spine (Phila Pa 1976)* **2016**, *41*, E139–E1147. [CrossRef]

10. Nagashima, H.; Dokai, T.; Hashiguchi, H.; Ishii, H.; Kameyama, Y.; Katae, Y.; Morio, Y.; Morishita, T.; Murata, M.; Nanjo, Y.; et al. Clinical features and surgical outcomes of cervical spondylotic myelopathy in patients aged 80 years or older: A multi-center retrospective study. *Eur. Spine J.* **2011**, *20*, 240–246. [CrossRef]

11. Yoshida, G.; Kanemura, T.; Ishikawa, Y.; Matsumoto, A.; Ito, Z.; Tauchi, R.; Muramoto, A.; Matsuyama, Y.; Ishiguro, N. The effects of surgery on locomotion in elderly patients with cervical spondylotic myelopathy. *Eur. Spine J.* **2013**, *22*, 2545–2551. [CrossRef]

12. Yamazaki, T.; Yanaka, K.; Sato, H.; Uemura, K.; Tsukada, A.; Nose, T. Cervical spondylotic myelopathy: Surgical results and factors affecting outcome with special reference to age differences. *Neurosurgery* **2003**, *52*, 122–126.

13. Son, D.K.; Son, D.W.; Song, G.S.; Lee, S.W. Effectiveness of the laminoplasty in the elderly patients with cervical spondylotic myelopathy. *Korean J. Spine* **2014**, *11*, 39–44. [CrossRef] [PubMed]

14. Kawaguchi, Y.; Kanamori, M.; Ishihara, H.; Ohmori, K.; Abe, Y.; Kimura, T. Pathomechanism of myelopathy and surgical results of laminoplasty in elderly patients with cervical spondylosis. *Spine (Phila Pa 1976)* **2003**, *28*, 2209–2214. [CrossRef] [PubMed]

15. Matsuda, Y.; Shibata, T.; Oki, S.; Kawatani, Y.; Mashima, N.; Oishi, H. Outcomes of surgical treatment for cervical myelopathy in patients more than 75 years of age. *Spine* **1999**, *24*, 529–534. [CrossRef] [PubMed]

16. Charlson, M.E.; Pompei, P.; Ales, K.L.; MacKenzie, C.R. A new method of classifying prognostic comorbidity in longitudinal studies: Development and validation. *J. Chronic Dis.* **1987**, *40*, 373–383. [CrossRef]

17. Benzel, E.C.; Lancon, J.; Kesterson, L.; Hadden, T. Cervical laminectomy and dentate ligament section for cervical spondylotic myelopathy. *J. Spinal Disord.* **1991**, *4*, 286–295. [CrossRef]

18. Hirabayashi, K.; Satomi, K. Operative procedure and results of expansive open-door laminoplasty. *Spine (Phila Pa 1976)* **1988**, *13*, 870–876. [CrossRef]

19. Tetreault, L.; Nouri, A.; Kopjar, B.; Cote, P.; Fehlings, M.G. The Minimum Clinically Important Difference of the Modified Japanese Orthopaedic Association Scale in Patients with Degenerative Cervical Myelopathy. *Spine (Phila Pa 1976)* **2015**, *40*, 1653–1659. [CrossRef]

20. Tetreault, L.A.; Karpova, A.; Fehlings, M.G. Predictors of outcome in patients with degenerative cervical spondylotic myelopathy undergoing surgical treatment: Results of a systematic review. *Eur. Spine J.* **2015**, *24* (Suppl. 2), 236–251. [CrossRef]

21. Holly, L.T.; Matz, P.G.; Anderson, P.A.; Groff, M.W.; Heary, R.F.; Kaiser, M.G.; Mummaneni, P.V.; Ryken, T.C.; Choudhri, T.F.; Vresilovic, E.J.; et al. Clinical prognostic indicators of surgical outcome in cervical spondylotic myelopathy. *J. Neurosurg. Spine* **2009**, *11*, 112–118. [CrossRef]

22. Tanaka, J.; Seki, N.; Tokimura, F.; Doi, K.; Inoue, S. Operative results of canal-expansive laminoplasty for cervical spondylotic myelopathy in elderly patients. *Spine (Phila Pa 1976)* **1999**, *24*, 2308–2312. [CrossRef]

23. Machino, M.; Yukawa, Y.; Hida, T.; Ito, K.; Nakashima, H.; Kanbara, S.; Morita, D.; Kato, F. Can elderly patients recover adequately after laminoplasty? A comparative study of 520 patients with cervical spondylotic myelopathy. *Spine (Phila Pa 1976)* **2012**, *37*, 667–671. [CrossRef] [PubMed]

24. Madhavan, K.; Chieng, L.O.; Foong, H.; Wang, M.Y. Surgical outcomes of elderly patients with cervical spondylotic myelopathy: A meta-analysis of studies reporting on 2868 patients. *Neurosurg. Focus* **2016**, *40*, E13. [CrossRef] [PubMed]

Journal of
Clinical Medicine

Article

Cervical Myelopathy in Patients Suffering from Rheumatoid Arthritis—A Case Series of 9 Patients and A Review of the Literature

Insa Janssen [1,2,*], Aria Nouri [1], Enrico Tessitore [1] and Bernhard Meyer [2,*]

1   Department of Neurosurgery, University of Geneva, 1205 Geneva, Switzerland; arianouri9@gmail.com (A.N.); Enrico.Tessitore@hcuge.ch (E.T.)
2   Department of Neurosurgery, Klinikum rechts der Isar, Technical University Munich, 81675 Munich, Germany
*   Correspondence: insajanssen@icloud.com (I.J.); bernhard.meyer@tum.de (B.M.)

Received: 1 February 2020; Accepted: 10 March 2020; Published: 17 March 2020

Abstract: Cervical myelopathy occurs in approximately 2.5% of patients suffering from rheumatoid arthritis (RA) and is associated with notable morbidity and mortality. However, the surgical management of patients affected by cervical involvement in the setting of RA remains challenging and not well studied. To address this, we conducted a retrospective analysis of our clinical database between May 2007 and April 2017, and report on nine patients suffering from cervical myelopathy due to RA. We included patients treated surgically for cervical myelopathy on the basis of diagnosed RA. Clinical findings, treatment and outcome were assessed and reported. In addition, we conducted a narrative review of the literature. Four patients were male. Mean age was 64.8 ± 20.5 years. Underlying cervical pathology was anterior atlantoaxial instability (AAI) associated with retrodental pannus in four cases, anterior atlantoaxial subluxation (AAS) in two cases and basilar invagination in three cases. All patients received surgical treatment via posterior fixation, and in addition two of these cases were combined with a transnasal approach. Preoperative modified Japanese orthopaedic association scale (mJOA) improved from 12 ± 2.4 to 14.6 ± 1.89 at a mean follow-up at 18.8 ± 23.3 months (range 3–60 months) in five patients. In four patients, no follow up was available, and the mJOA of these patients at time of discharge was stable compared to the preoperative score. One patient died two days after surgery, where a pulmonary embolism was assumed to be the cause of mortality, and one patient sustained a temporary worsening of his neurological deficit postoperatively. Surgery is generally an effective treatment method in patients with inflammatory arthropathies of the cervical spine. Given the nature of the RA and potential instability, fixation in addition to cord decompression is generally required.

Keywords: cervical myelopathy; spinal cord compression; cervical spine surgery; rheumatoid arthritis (RA); cranial settling (CS); atlantoaxial subluxation (AAS); atlantoaxial instability (AAI)

---

1. Introduction

Cervical myelopathy occurs in approximately 2.5% of patients suffering from rheumatoid arthritis (RA) for more than 14 years and is associated with notable morbidity and mortality [1]. In these cases, compression of the spinal cord and brain stem is typically caused by atlantoaxial instability (AAI) or atlantoaxial subluxation (AAS) with associated vertical migration of the dens of C2 and accompanying retrodental pannus formation [2].

Rheumatoid arthritis is a chronic inflammatory systemic disease and affects about 1–2% of adults [3]. Besides peripheral joints, the upper cervical spine is preferentially afflicted in patients with RA due to the anatomic vulnerably of numerous synovial and apophyseal joints to dynamic forces [4,5].

This is especially true for the cranio-cervical junction and the atlantoaxial joints. In these patients, an inflammatory synovial proliferation results in deterioration of ligamentous tendinous tissues and bone erosion, thereby increasing sliding motion between the atlas and axis, and as a consequence, atlantoaxial instability (AAI) [5]. In the advanced stage, an atlantoaxial subluxation (AAS) can develop and manifest in the vertical migration of the dens, known as cranial settling (CS), basilar impression, basilar invagination or vertical translocation (VT) [4,6]. Most common is an anterior atlantoaxial subluxation due to an affected anterior median atlanto-axial joint between ventral arch of C1 and the dens of the axis (circa 75% of all AAS). Subluxation in the posterior median atlanto-axial joint, located between posterior arch of C1 and the dens of the axis, as well as asymmetrical or unilateral changes of the lateral atlanto-axial joint leading to impairment in rotation are rare (7% and 20%) [7]. For diagnosis of cervical involvement in RA various diagnostic imaging options are available, each playing a different role depending on the stage of involvement (see discussion, Figures 1 and 2, Table 1).

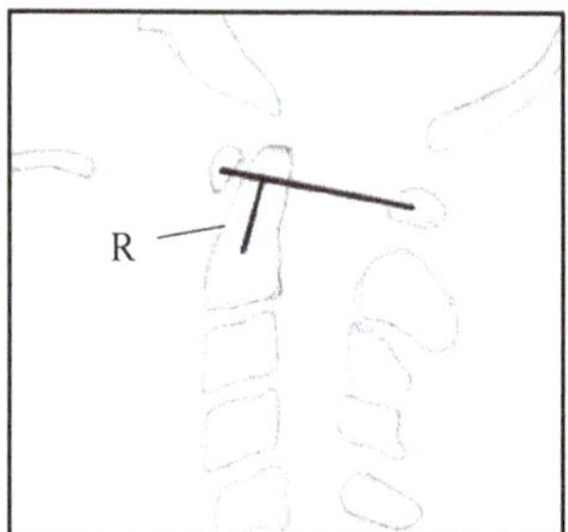
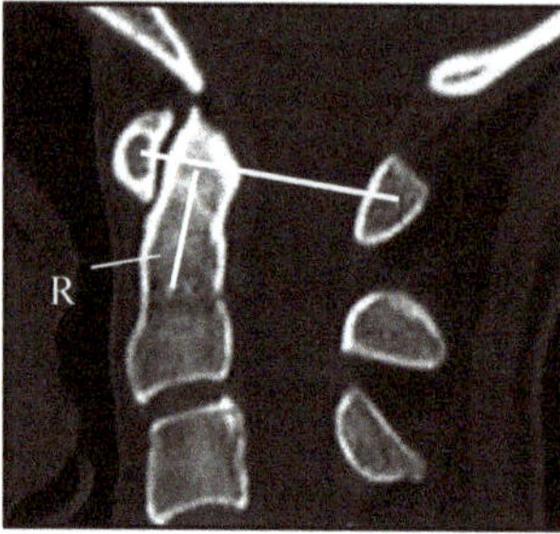
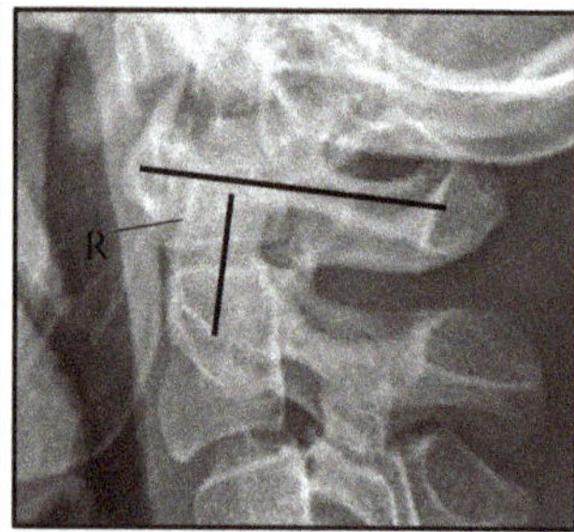

Figure 1. Ranawat criteria: Line from the center of the anterior C1 arch to the center of the posterior C1 arch. A second line goes through the axis from the odontoid to the center of the base of the C2. The smaller the distance (R) to the crossing of the lines, the more pronounced the invagination. A distance > 13 mm for women and > 15 mm for men is normal [8].

Table 1. Definition and diagnostics of cervical instabilities.

Type of Instability	Definition and Diagnostics of Cervical Instabilities	
	Definition	Diagnostic in Radiograph/Scan in Lateral/Sagittal Projection
AAS (atlantoaxial subluxation)	weakening or rupture of ligaments and subchondral bone erosion in the atlantoaxial joints	anterior atlantodental Interval (AADI) > 3 mm posterior atlantodental interval (PADI) < 14 mm
SAS (subaxial subluxation)	subluxation in the joints C3-7 due to destruction of the joint surface and the ligaments between the processes spinosis	horizontal displacement of vertebrae with irreducible translation > 3.5 mm
CS (Cranial Settling)	vertical translocation of dens into the foramen magnum	see Figures 1 and 2

After the introduction of early and consistent therapy with biologicals, the prevalence of cervical involvement in RA has been declining and is estimated to be between 20–44%, whereas prevalence rates in historical data present estimates between 42–80% [9]. Symptomatic involvement of the cervical spine in terms of instability manifests itself on average about 7–10 years after initial diagnosis of RA [3,10]. Atlantoaxial subluxation with myelopathy is seen in approximately 2.5% of patients with RA for more than 14 years [5]. For prevalence and symptoms, see Table 2. Medical therapy of RA at the early stages is feasible to prevent cervical involvement in RA. Early diagnosis and thus an early start of a sufficient drug therapy is regarded as a prognostically positive factor with regard to a risk reduction for a manifestation of cervical involvement [11]. Thanks to drug therapies, the instabilities requiring surgery are statistically decreasing, however, the cases needing treatment are becoming more complex [12].

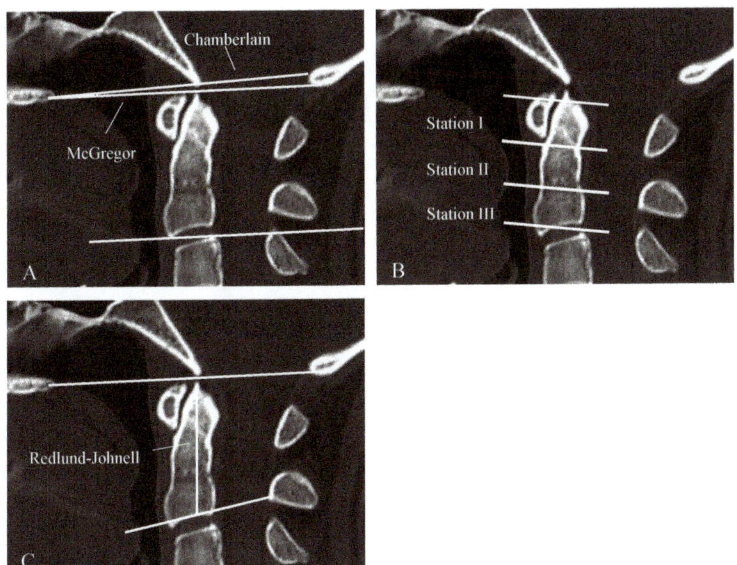

Figure 2. Various diagnostic criteria of cranial settling. Chamberlaine: Line from the dorsal end of the hard palate to the posterior border of the foramen magnum. If the dens is 3 mm above the line, a basilar invagination is present by definition (**A**). McGregor: Line from the hard palate to the deepest point of the occiput. A basilar invagination is present if the tip of the dens is more than 4.5 mm above the line (**A**). Clark-Station: Here the dens is divided into three equally sized stations. If the front arch of the atlas reaches into station II or III the criteria of a cranial settling is fulfilled (**B**). Redlund-Johnell: describes the distance between the center of the lower cover plate C2 and the McGregor line. Normal is 34 mm for men and 29 mm for women (**C**).

Table 2. Prevalence and symptoms of pat with clinical involvement [6].

Prevalence of Cervical Involvement in RA	%
Pain in the cranio-cervical junction	69% of patients with cervical instability
Muscular atrophy, paresis, bladder rectal disorders, pathological reflexes and spasticity	present in up to 58% of all cases
Involvement of the cranial nerves	reported in about 20%
Initially asymptomatic	33–50%
Atlantoaxial subluxation with myelopathy	circa 2.5% of patients with RA for more than 14 years
Locked-in syndrome or sudden death	rare but reported up to 10% in a postmortem study [13]
Vertebrobasilar insufficiency with tinnitus and dizziness due to Mechanical compression/ vertebrobasilar thromboembolic events due to kinking of vertebral arteries	rare
Aseptic discitis and atraumatic dens fractures	rare

While conservative therapy can effectively palliate pain, operative treatment is indicated in medically therapy resistant pain with radiographic instability, and is necessary after the development of neurological deficit, myelopathy, cranial nerve and bulbar dysfunction [8].

The objective of the present paper is to display the variety of disorders that present and discuss this in the management of such cases in the context of current literature.

2. Methods

A retrospective analysis of the clinical database of the department of neurosurgery in Munich was performed between May 2007 and April 2017, with research ethics board approval (Ethics committee of Technical University of Munich, No.238/17 S).

We searched the database for patients which had received surgery for craniocervical and subaxial instabilities. Patients with traumatic or neoplastic retro-odontoid pannus or craniocervical junction abnormalities like chiari malformation were not included in this series. Out of this cohort, we investigated patients treated surgically for cervical myelopathy on the basis of diagnosed or suspected rheumatoid arthritis or chronic inflammatory systemic disease. We excluded patients with pure degenerative instabilities. Clinical findings, treatment and outcome were assessed. In addition, we conducted a review of the literature. The aim of the study was to evaluate variations in patient characteristics, surgical treatment and patient outcomes. To assess the cervical involvement and severity of myelopathy, the Ranawat classification (Table 3) and criteria and the modified Japanese orthopaedic association scale (mJOA); modified by Keller, 1993) were used to assess degree of neurological function [7,14]. Post-operative mJOA was obtained at the last follow-up, at least 3 months following operation.

Table 3. Ranawat classification [15].

Class	Description
I	Pain, no neurological deficit
II	Subjective weakness, hyperreflexia, dysesthesia
III	Objective weakness, long-tract signs III A—Ambulatory, III B—Non-ambulatory

3. Results

In the defined period of time, 39 patients received surgery for rheumatic and degenerative cervical instabilities. Six patients were known for RA and under medical treatment. In five patients a chronic inflammatory systemic disease was suspected but not diagnosed at time of neurosurgical treatment. Out of 39 patients, myelopathy was present in 16 patients (47.1%).

Out of eleven patients known for rheumatoid arthritis (RA) or suspected chronic inflammatory systemic disease, 9 patients were suffering from myelopathy. Four patients were male, five females. Mean age was 64.8 ± 20.5 years (range 22–82 years). All patients presented with clinical myelopathy, and additionally some patients presented with further neurological deficits (see Tables 4 and 5). Surgeries were performed by seven surgeons. The underlying cervical pathology was AAI of the median atlanto-axial joint combined with retrodental pannus in four, anterior AAS in two and basilar invagination in three cases. Four patients were already known for RA and under medical treatment. In five patients a chronic inflammatory systemic disease was suspected but not diagnosed at time of neurosurgical treatment. The diagnosis for RA was confirmed later by a rheumatologist in each case. Patients were classified according to the Ranawat criteria (see Tables 3 and 5).

Table 4. Case description.

Case	Case 1	Case 2	Case 3	Case 4	Case 5	Case 6	Case 7	Case 8	Case 9
Sex/age	m/80	w/22	w/82	m/76	m/73	w/71	m/48	w/81	w/50
Cervical pathology	AAI, retrodental pannus	AAS	AAI, retrodental pannus	CS	AAI, retrodental pannus, cervical spinal stenosis subaxial	AAI, retrodental pannus	AAS, retrodental pannus	CS	CS
Pre-operative mJOA - score	9	12	15	8	15	16	14	11	14
Ranawat - criteria	15.5	10.7	12	8.3	12.1	*	13.5	9.5	*
Smallest diameter of canal anterior - posterior (mm)	1.9	4	5.4	5.9	7.2	*	5.1	5.5	*
ADI (mm)	0.2	9.7	0.3	1.3	1.4	*	13.0	0.8	*
Ranawat classification	IIIA	IIIA	IIIA	IIIB	IIIA	II	II	IIIB	IIIA
Rheumatoid arthritis/treatment	suspicion of RA	RA/Adalimumab every 2 weeks	suspicion of RA	RA/MTX every week, folic acid, steroids	suspicion of RA	suspicion of RA	RA (ED 1995)/MTX 1 every week, steroids	RA since 30 years/steroids	suspicion of RA
Symptoms	neck pain, ataxia, hemiparesie, not able to walk	fine motor disorders, monoparesis, bladder emptying disorders, ataxia	monoparesis	tertaparesis, not able to walk, dysphagia, loss of warm cold discrimination of the legs	monoparesie, ataxie	neck pain, ataxie	ataxie, sensory deficit	dysphagia, tetraparesie, not able to walk	ataxie, dysphagia, sensory deficit
Surgery	1. posterior fixation C1-3 + laminectomy C1-2 2. transnasal endoscopic dens resection	closed reduction, osterior fixation C1-2	posterior fixation C1-2, laminectomy C1	posterior fixation C0-2-3-4	posterior fixation C1-2-4-6 + laminectomy C1-6	posterior fixation C1-2	posterior fixation C1-2	posterior fixation C0-3-4, laminectomy C1, decompression suboccipital	1. posterior fixation C0-2, decompression suboccipital 2. transnasal endoscopic resection of dens and clivus (21 months later)
Complications	temporary hemiplegie postoperatively	no	no	deceased	No	no	no	no	no
Follow up	after 8 months, improvement of hemipaesis, walking possible with aide	after 12 months, no symptoms	no follow up	no follow up	no follow up	no follow up	11 months	after 3 months, walking possible	after 5 years
Post-operative mJOA score	12	17	15	deceased	15	16	16	12	16

* Preoperative images not available.

133

Table 5. Demographics.

Number of Patients/Sex	$n = 9$ (Female $n = 5$; Male $n = 4$)
Mean age	64.8 ± 20.5 years (range 22–82 years)
Cervical pathology	AAI with retrodental pannus, $n = 4$ Anterior AAS, $n = 2$ Basilar invagination, $n = 3$ Associated subaxial spinal stenosis $n = 2$
Myelopathy	$n = 9$
Additional neurological deficit	Tetraparesis ($n = 2$) Hemi-or monoparesis ($n = 4$) Incontinence ($n = 1$) Pain ($n = 4$) Dysphagia ($n = 3$)
Pre-operative mJOA- score (mean)	12.67 ± 2.83 (range 8–16)
Ranawat classification	Class II $n = 2$ Class IIIA $n = 5$ Class IIIB $n = 2$
Surgery	posterior fixation, $n = 9$ C1-2, $n = 4$; C1-6, $n = 1$ C1-3, $n = 1$; C0-2, $n = 1$ C0-3, $n = 1$; C0-4, $n = 1$ posterior decompression, $n = 6$ transnasal endoscopic dens resection, $n = 2$
Follow up	mean follow-up at 18.8 ± 23.3 months (range 3–60 months)
Post-operative mJOA score (mean) *	14.6 ± 1.89 (range 12–17)
Mortality	11.1% ($n = 1$)

* evaluated in 5 patients.

Mean preoperative modified Japanese orthopaedic association scale (mJOA) at admission was 12.67 ± 2.83 (range 8-16). All patients were suffering from symptoms for several month and sustaining a progressive worsening within few weeks before surgical treatment. Five patients were observed with a mean follow-up at 18.8 ± 23.3 months (range 3–60 months), all of which showed improvement of pain and neurological deficits. The mean preoperative mJOA score in these patients showed an improvement from 12 ± 2.4 to 14.6 ± 1.89, but only one patient made a complete recovery (case 2). In four patients, no follow up was available, and the mJOA of these patients at time of discharge was stable compared to the preoperative score.

Surgical treatment via posterior fixation was conducted in all cases. In the presence of AAI (n = 2) and AAS ($n = 2$) associated with retrodental pannus, patients underwent a posterior C1-C2 stabilization using the Goel-Harm's technique. Due to an associated subaxial spinal stenosis two patients underwent a cervical fixation from C1 to C3 and C1 to C6. In patients of underlying basilar invagination, stabilization including C0 was performed ($n = 3$). Posterior decompression was necessary in six cases.

In two cases, a ventral decompression in terms of a transnasal endoscopic dens resection was performed, whereas in one case the operation took place two years after the first intervention (case 9). A 50- year old patient first showed an improvement of myelopathy after posterior fixation C0-2 and suboccipital de-compression for basilar invagination, but she again sustained a worsening of symptoms after two years. The other patient was diagnosed for AAI with retrodental pannus.

One patient suffering from basilar invagination resulting in tetraparesis (Ranawat class IIIB) died on the second postoperative day. The cause of death remained unclear, but was suspected to be due a pulmonary embolism. In one patient (Ranawat class IIIA) a temporary worsening of the neurological deficit after posterior fixation for AAI and retrodental pannus occurred. He received a second surgery for transnasal ventral decompression and mainly recovered from his hemiparesis. Among the other patients, there was no complication, and none showed a new neurological deficit (Tables 4 and 5).

Illustrative Case (2)

A 22-year-old female patient presented at emergency with rapidly progressive fine motor disorders and paresis of the right arm. On clinical examination, mild urinary retention as well as ataxia was also observed, and she scored 12 on the mJOA scale. The patient was known for RA for several years and was treated with Adalimumab every two weeks. Imaging showed an atlantodental instability with retrodental pannus formation and basilar invagination. Basilar invagination was confirmed by a Ranawat criteria of 10.7 mm and an anterior atlantodental interval of 9.7 mm, indicating a Ranawat IIIA classification. The first step was a closed reduction under X-ray control and the application of a rigid neck brace on the day of admission in order to treat the symptoms immediately. In the second step, open reduction and fixation via implantation of an internal fixator took place two days later.

The nerve root C2 was exposed and rhizotomized on both sides, the facet joint between C1/2 distracted and autologous bone chips inserted. After three months, residual sensory deficit of the fingertips on both sides remained, a motor deficit, urinary retention and ataxia were no longer present. There was no relevant pain symptomatology. After 12 months, she had made a complete recovery, and scored 17 on the mJOA scale—see Figure 3.

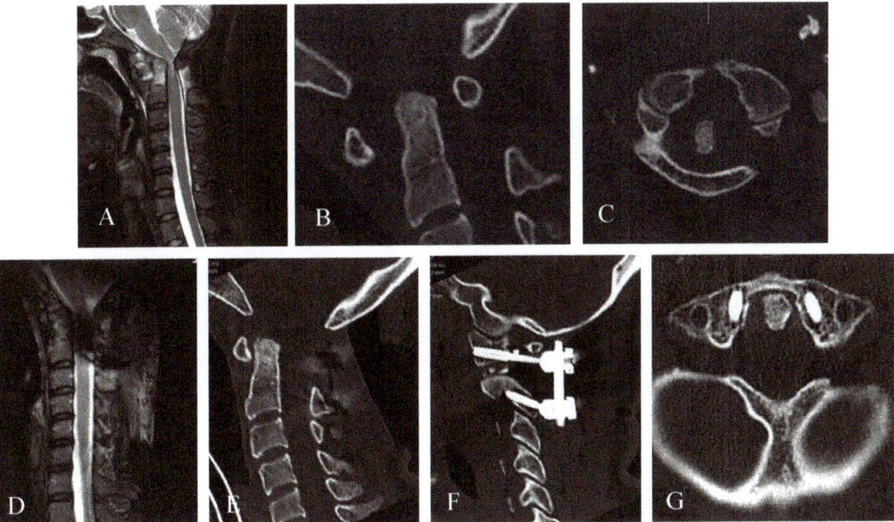

Figure 3. (**A**) Preoperative MRI imaging of the cervical spine atlantodental instability with retrodental pannus formation and basilar impression due to dorsal displacement of the dens resulting in absolute spinal constriction and myelopathy. (**B,C**) Enlarged anterior atlantodental interval in the preoperative scan (AADI). (**D–G**) Postoperative imaging after initially closed and then open reduction and fixation C1-2.

4. Discussion

Our results show that surgical treatment is a very effective treatment method in patients with inflammatory arthropathies of the cervical spine. The patients available for follow up showed an improvement of preoperative mJOA from 12 ± 2.4 to 14.6 ± 1.89, the others showed a stable score at time of discharge without a new neurological deficit. However, in our case series two patients sustained a relevant complication, with one patient dying the second postoperative day (likely due to a pulmonary embolism) and the other experiencing a temporary worsening of a hemiparesis after posterior fixation for AAI and retrodental pannus was achieved. The latter patient received a second surgery via transnasal ventral decompression and mainly recovered from his hemiparesis in the following months. Both of these patients presented with a severe neurological deficit (preoperative mJOA 8 and 9, Ranawat class IIIA and B) which confirms the assumption that perioperative morbidity and mortality increases significantly in rheumatoid non ambulatory (Ranawat Class IIIB) patients with loss of the ability to walk [8,16].

In case of isolated AAI without subluxation C1-C2 fixation associated with (n = 5) and without (n = 4) C1 laminectomy was performed. In one case disappearance of the pseudotumor failed, and a ventral approach was necessary after two years, which was conducted endoscopically via a transnasal approach.

Pain in the craniocervical junction is a typical symptom that is seen in 69% of patients with cervical instability [3]. Occipital neuralgia, facial hypoesthesia, facial pain as well as pain in the mastoid area and ear pain may also manifest as they are caused by irritation of the occipital nerve between C1 and C2 or by irritation of the trigeminal nucleus in the medulla oblongata [3]. Patients can also suffer from radiculopathy of the upper extremities. Neural deficits like myelopathy, muscular atrophy, paresis, bladder rectal disorders, pathological reflexes and spasticity are present in up to 58% of all cases [6,17,18]. Compression of the medulla oblongata can also lead to disorders of the caudal cranial nerves such as dysphagia and dysarthria resulting from compression of the vagus, glossopharyngeal and hypoglossal nerve. Involvement of the cranial nerves is reported in about 20% of cases [3]. Beyond the development of severe neurological symptoms, locked-in syndrome or sudden death can also occur [17,19]. Mechanical compression can cause the development of a vertebrobasilar insufficiency with tinnitus and dizziness. Similarly, instability can cause a kinking of vertebral arteries, which can lead to vertebrobasilar thromboembolic events. Another, but less common consequence of RA, is the destruction and instability of the subaxial spine resulting in subaxial dislocation (SAS) [3,6,17]. Aseptic discitis and atraumatic dens fractures have also been reported and can also be the consequence of the RA inflammatory process [3]. For prevalence and symptoms, see also Table 1.

Traditional imaging is conducted using conventional x-ray of the cervical spine, which gives information about spinal alignment. As it is related to lower cost, it is widely available and has a low radiation dose, it is especially suitable for screening of asymptomatic patients with RA in the outpatient setting [5,6]. Additionally, flexion- extension images allow for easy visualization of occult instabilities [6]. However, the gold standard for bone evaluation and evaluation of ankylosis and pseudarthrosis is the cervical CT scan, which is also often acquired for surgical planning. For the assessment of soft tissue, spinal cord, or nerve compression and myelopathy, cervical MRI is the imaging modality of choice. It is always indicated for the evaluation of patients with neurology deficit [6].

The conventional x-ray of the cervical spine is conducted in anterior-posterior and lateral projection, supplemented by flexion-extension images as well as radiographs through the open mouth aimed at the odontoid process [5,6]. In case of abnormalities in conventional images, clinical suspicion or neural deficits, cervical CT and MRI are added. By definition, an atlantoaxial subluxation is present if the atlantodental interval (AADI) measured on lateral radiographs is >3 mm. Evidence of AAS is also shown by a distance between the dens and C1 posterior arch (posterior atlantodental interval (PADI) <14 mm [7].

For the assessment of vertical subluxation (VS) a number of methods have been developed to describe the degree of dens displacement with regard to foramen magnum. One popular diagnostic method is the Ranawat criteria, which is based on a line which connects the center of the anterior arch with the center of the posterior arch of C1 vertebra and a second line which is drawn along the axis of the odontoid process, from the center of the base of C2 vertebra to the intersection with the first line (Figure 1). The distance is used to assess the presence of the basilar invagination and based on the criterion are defined to be present when < 13 mm in women and < 15 mm in men [7]. Other diagnostic criteria of cranial settling are described in Figure 2. SAS is diagnosed when the radiograph in lateral projection shows horizontal displacement of vertebrae with irreducible translation by >3.5 mm (Table 1) [7].

Besides the grade of AAS, the localization and the extend of neurological impairment, atlanto-axial instability can be classified as reducible, partially reducible or fixed, according to the response to traction [4]. RA with cervical involvement is a progressive and serious condition with reduced lifetime expectancy. Matsunaga et al. noted that all of the irreducible AAS patients who did not undergo surgical treatment were bedridden within 3 years after the onset of myelopathy, and the survival rate was 0% in the first eight years [15,20]. Atlantodental involvement is typically a late manifestation of RA [3]. However, patients with a worse disease activity score, high disease activity and erosive disease at baseline have a high risk of atlantoaxial involvement in early rheumatoid arthritis (ERA) (disease duration < 12 months) [1]. Affection of the small joints of the hand and foot, failure of antirheumatic therapies, intake of glucocorticoids, young age at diagnosis, level of CRP, female sex, and low BMI were identified as risk factors for cervical involvement [21,22].

In the early stage, clinical involvement is initially asymptomatic in 33–50% of patients, which makes early drug intervention before the development of instability more difficult. Symptoms of cervical myelopathy are often misinterpreted due to the rheumatic affection of the finger and ankle joints, which may mask myelopathy as the cause of fine motor disorders and gait insecurity.

In case of isolated AAS without vertical instability or relevant destruction of the C0-C1 joints, atlantoaxial fusion via C1-C2 fixation is the means of choice [17]. The stabilization should only be extended to C0 if not otherwise possible, as this means an important limitation of flexion and extension [23]. The atlanto-axial transarticular screw fixation introduced by Magerl has been the gold standard for treatment of atlanto- axial instability for long time [24]. At present, due to reduced risk for neurovascular complications, the surgical technique of choice is the technique described by Harms and Goel, containing C1 lateral mass and C2 pedicle screws which has proven its efficacy and safety [25]. In comparison to transarticular screw fixation, the Harms technique also includes reduction of C1-C2 and the protection of the C1-C2 joint [15].

Direct removal of intracanalar tissue compressing the spinal cord is not obligatory [17]. After fixation of an instability by posterior stabilization, the retrodental pannus usually recedes [2].

Anterior decompression may be indicated for cases that are irreducible or in cases in which a pannus fails to regress after posterior fixation. By distraction of facet C1-C2 an improvement of the cranial setting is possible [17].

The transoral dens resection is known as the classical method for a ventral decompression, but it is associated with considerable risks. In 33.6% of cases the splitting of the soft palate is necessary, resulting complications include hypernasal speech, nasal regurgitation, tracheal edema and fistula formation in the pharynx. As a result, 14.8% of patients require a PEG and 3.8% a tracheotomy. A velopharyngeal insufficiency results in 4% of cases, and the mortality rate is 2.3% [26,27]. In a case series by Gempt et al. the transnasal endoscopic approach could be shown as an alternative to transoral decompression [28]. If a ventral dens resection is necessary, the endoscopic transnasal approach is preferred. Compared to the transoral approach, the patients can be extubated after 0.3 days instead of 3.5 days after transnasal endoscopic access, and oral nutrition can be started after one day instead of 5.2 days [26,27].

Given the fact that a relevant number of patients are asymptomatic in the stage of cervical involvement, the objective beside the optimal surgical management should be the early detection of cervical involvement. In a study by Neva et al., a previously undiagnosed cervical subluxation could be detected in 44% of a total of 194 patients included who underwent an orthopaedic surgery of the peripheral joints due to RA (Mean age was 65 years, mean duration of RA was 24 years). Compared to those without subluxation, the affected patients did not show a more frequent presence of symptoms such as neck pain, occipital, temporal, retro-orbital pain or radiculopathy of the upper extremities. Thus, by clinical appearance alone, patients with subluxation are not distinguishable from those without [29]. Therefore, a radiological screening of patients with RA should be considered.

5. Conclusions

Surgical treatment is an effective treatment method in patients with inflammatory arthropathies of the cervical spine, whereas the risk of perioperative morbidity and mortality increases significantly with the severity of neurological impairment. Due to the advent of disease modifying medication, the prevalence of cervical involvement in RA seems to be diminishing. However, when present, these cases can be challenging to treat. Given the dynamic nature of RA myelopathy, surgical treatment generally requires fixation and was performed in all patients in the presented cohort.

Author Contributions: Conceptualization, I.J. methodology, I.J., validation, I.J., investigation, I.J.; resources, I.J., data curation, I.J., writing—original draft preparation, I.J.; writing—review and editing, B.M., A.N. and E.T.; visualization, I.J.; supervision, E.T. All authors have read and agreed to the published version of the manuscript.

Conflicts of Interest: Insa Janssen and Aria Nouri declare no conflict of interest. Enrico Tessitore declares receiving training fees from Spineart, NuVasive and DePuy Synthes. Bernhard Meyer declares receiving fees for advisory board from Spineart, DePuy Synthes, Ulrich Medical, Brainlab and Relivant. The funders had no role in the design of the study; in the collection, analyses, or interpretation of data; in the writing of the manuscript, or in the decision to publish the results.

Abbreviations

RA	rheumatoid arthritis
CS	cranial settling
ASS	atlantoaxial subluxation
AAI	atlantoaxial instability
mJOA	modified Japanese orthopaedic association scale
SAS	subaxial subluxation
AADI	anterior atlantodental Interval
PADI	posterior atlantodental interval
VT	vertical translocation
CT	computer tomography
MRI	magnetic resonance tomography

References

1. Carotti, M.; Salaffi, F.; di Carlo, M.; Sessa, F.; Giovagnoni, A. Magnetic resonance imaging of the craniovertebral junction in early rheumatoid arthritis. *Skelet. Radiol.* **2019**, *48*, 553–561. [CrossRef] [PubMed]

2. Certo, F.; Maione, M.; Visocchi, M.; Barbagallo, G.M.V. Retro-odontoid Degenerative Pseudotumour Causing Spinal Cord Compression and Myelopathy: Current Evidence on the Role of Posterior C1-C2 Fixation in Treatment. *Acta Neurochir. Suppl.* **2019**, *125*, 259–264. [PubMed]

3. Gillick, J.L.; Wainwright, J.; Das, K. Rheumatoid Arthritis and the Cervical Spine: A Review on the Role of Surgery. *Int. J. Rheumatol.* **2015**, *2015*, 252456. [CrossRef] [PubMed]

4. Joaquim, A.F.; Appenzeller, S. Cervical spine involvement in rheumatoid arthritis—A systematic review. *Autoimmun. Rev.* **2014**, *13*, 1195–1202. [CrossRef] [PubMed]

5. Wasserman, B.R.; Moskovich, R.; Razi, A.E. Rheumatoid arthritis of the cervical spine—Clinical considerations. *Bull. NYU Hosp. Joint. Dis.* **2011**, *69*, 136–148.

6. Joaquim, A.F.; Ghizoni, E.; Tedeschi, H.; Appenzeller, S.; Riew, K.D. Radiological evaluation of cervical spine involvement in rheumatoid arthritis. *Neurosurg. Focus* **2015**, *38*, E4. [CrossRef]

7. Manczak, M.; Gasik, R. Cervical spine instability in the course of rheumatoid arthritis—Imaging methods. *Reumatologia* **2017**, *55*, 201–207. [CrossRef]

8. Sunahara, N.; Matsunaga, S.; Mori, T.; Ijiri, K.; Sakou, T. Clinical course of conservatively managed rheumatoid arthritis patients with myelopathy. *Spine* **1997**, *22*, 2603–2607; discussion 2608. [CrossRef]

9. Neva, M.H.; Kaarela, K.; Kauppi, M. Prevalence of radiological changes in the cervical spine—A cross sectional study after 20 years from presentation of rheumatoid arthritis. *J. Rheumatol.* **2000**, *27*, 90–93.

10. Del Grande, M.; del Grande, F.; Carrino, J.; Bingham, C.O., 3rd; Louie, G.H. Cervical spine involvement early in the course of rheumatoid arthritis. *Semin. Arthritis Rheum.* **2014**, *43*, 738–744. [CrossRef]

11. Kothe, R.; Wiesner, L.; Ruther, W. Rheumatoid arthritis of the cervical spine. Current concepts for diagnosis and therapy. *Orthopade* **2002**, *31*, 1114–1122. [CrossRef] [PubMed]

12. Stein, B.E.; Hassanzadeh, H.; Jain, A.; Lemma, M.A.; Cohen, D.B.; Kebaish, K.M. Changing trends in cervical spine fusions in patients with rheumatoid arthritis. *Spine* **2014**, *39*, 1178–1182. [CrossRef] [PubMed]

13. Mikulowski, P.; Wollheim, F.A.; Rotmil, P.; Olsen, I. Sudden death in rheumatoid arthritis with atlanto-axial dislocation. *Acta Med. Scand.* **1975**, *198*, 445–451. [CrossRef] [PubMed]

14. Keller, A.; von Ammon, K.; Klaiber, R.; Waespe, W. Spondylogenic cervical myelopathy: Conservative and surgical therapy. *Schweiz. Med. Wochenschr.* **1993**, *123*, 1682–1691. [PubMed]

15. Bhatti, A.B.; Kim, S. Application of the Harms Technique to Treat Undiagnosed Intractable C1-C2 Unilateral Neck Pain: A Case Report. *Cureus* **2016**, *8*, e793. [CrossRef] [PubMed]

16. Nannapaneni, R.; Behari, S.; Todd, N.V. Surgical outcome in rheumatoid Ranawat Class IIIb myelopathy. *Neurosurgery* **2005**, *56*, 706–715; discussion 706–715. [CrossRef] [PubMed]

17. Ryu, J.I.; Han, M.H.; Cheong, J.H.; Kim, J.M.; Kim, C.H.; Chun, H.J.; HumBak, K. The Effects of Clinical Factors and Retro-Odontoid Soft Tissue Thickness on Atlantoaxial Instability in Patients with Rheumatoid Arthritis. *World Neurosurg.* **2017**, *103*, 364–370. [CrossRef]

18. Nguyen, H.V.; Ludwig, S.C.; Silber, J.; Gelb, D.E.; Anderson, P.A.; Frank, L.; Vaccaro, A.R. Rheumatoid arthritis of the cervical spine. *Spine J.* **2004**, *4*, 329–334. [CrossRef]

19. Dohzono, S.; Suzuki, A.; Koike, T.; Takahashi, S.; Yamada, K.; Yasuda, H.; Yasuda, H.; Nakamura, H. Factors associated with retro-odontoid soft-tissue thickness in rheumatoid arthritis. *J. Neurosurg. Spine* **2016**, *25*, 580–585. [CrossRef]

20. Matsunaga, S.; Sakou, T.; Onishi, T.; Hayashi, K.; Taketomi, E.; Sunahara, N.; Komiya, S. Prognosis of patients with upper cervical lesions caused by rheumatoid arthritis: Comparison of occipitocervical fusion between c1 laminectomy and nonsurgical management. *Spine* **2003**, *28*, 1581–1587; discussion 1587. [CrossRef]

21. Terashima, Y.; Yurube, T.; Hirata, H.; Sugiyama, D.; Sumi, M.; Spinal, D.H.O. Predictive Risk Factors of Cervical Spine Instabilities in Rheumatoid Arthritis: A Prospective Multicenter Over 10-Year Cohort Study. *Spine* **2017**, *42*, 556–564. [CrossRef]

22. Zhu, S.; Xu, W.; Luo, Y.; Zhao, Y.; Liu, Y. Cervical spine involvement risk factors in rheumatoid arthritis: A meta-analysis. *Int. J. Rheum. Dis.* **2017**, *20*, 541–549. [CrossRef]

23. Werle, S.; Ezzati, A.; ElSaghir, H.; Boehm, H. Is inclusion of the occiput necessary in fusion for C1-2 instability in rheumatoid arthritis? *J. Neurosurg. Spine* **2013**, *18*, 50–56. [CrossRef]

24. Grob, D.; Jeanneret, B.; Aebi, M.; Markwalder, T.M. Atlanto-axial fusion with transarticular screw fixation. *J. Bone Joint. Surg. Br.* **1991**, *73*, 972–976. [CrossRef]

25. Lee, S.H.; Kim, E.S.; Sung, J.K.; Park, Y.M.; Eoh, W. Clinical and radiological comparison of treatment of atlantoaxial instability by posterior C1-C2 transarticular screw fixation or C1 lateral mass-C2 pedicle screw fixation. *J. Clin. Neurosci.* **2010**, *17*, 886–892. [CrossRef]

26. Goldschlager, T.; Hartl, R.; Greenfield, J.P.; Anand, V.K.; Schwartz, T.H. The endoscopic endonasal approach to the odontoid and its impact on early extubation and feeding. *J. Neurosurg.* **2015**, *122*, 511–518. [CrossRef]

27. Komotar, R.J.; Starke, R.M.; Raper, D.M.; Anand, V.K.; Schwartz, T.H. Endoscopic endonasal compared with anterior craniofacial and combined cranionasal resection of esthesioneuroblastomas. *World Neurosurg.* **2013**, *80*, 148–159. [CrossRef]

28. Gempt, J.; Lehmberg, J.; Grams, A.E.; Berends, L.; Meyer, B.; Stoffel, M. Endoscopic transnasal resection of the odontoid: Case series and clinical course. *Eur. Spine J.* **2011**, *20*, 661–666. [CrossRef]

29. Neva, M.H.; Hakkinen, A.; Makinen, H.; Hannonen, P.; Kauppi, M.; Sokka, T. High prevalence of asymptomatic cervical spine subluxation in patients with rheumatoid arthritis waiting for orthopaedic surgery. *Ann. Rheum. Dis.* **2006**, *65*, 884–888. [CrossRef]

Journal of
Clinical Medicine

Article

Quantitative Assessment of Gait Characteristics in Degenerative Cervical Myelopathy: A Prospective Clinical Study

Sukhvinder Kalsi-Ryan [1,2,3,4,†], Anna C. Rienmueller [3,4,5,†], Lauren Riehm [3], Colin Chan [3], Daniel Jin [6], Allan R. Martin [3,4], Jetan H. Badhiwala [3,4], Muhammad A. Akbar [3,4], Eric M. Massicotte [3,4] and Michael G. Fehlings [3,4,*]

1   KITE-UHN, Toronto, ON M5G 2A2, Canada; Sukhvinder.Kalsi-Ryan@uhn.ca
2   Department of Physical Therapy, University of Toronto, Toronto, ON M5G 1V7, Canada
3   Spine Program; Krembil Brain Institute; University Health Network, Toronto, ON M5T 2S8, Canada;
    anna.rienmuller@mail.utoronto.ca (A.C.R.); lauren.riehm@medportal.ca (L.R.); cchan827@uwo.ca (C.C.);
    allan.martin@mail.utoronto.ca (A.R.M.); jetan.badhiwala@mail.utoronto.ca (J.H.B.);
    muhammad.akbar@mail.utoronto.ca (M.A.A.); Eric.Massicotte@uhn.ca (E.M.M.)
4   Department of Surgery and Spine Program, University of Toronto, Toronto, ON M5T 1P5, Canada
5   Department of Orthopedic Surgery and Traumatology, Medical University Vienna, 1090 Vienna, Austria
6   Department of Kinesiology, University of Waterloo, Waterloo, ON N2L 3G1, Canada;
    drqjin@edu.uwaterloo.ca
*   Correspondence: Michael.fehlings@uhn.ca
†   These authors contributed equally to this work.

Received: 1 February 2020; Accepted: 4 March 2020; Published: 10 March 2020

Abstract: It is challenging to discriminate the early presentation of Degenerative Cervical Myelopathy (DCM) as well as sensitively and accurately distinguishing between mild, moderate, and severe levels of impairment. As gait dysfunction is one of the cardinal symptoms of DCM, we hypothesized that spatiotemporal gait parameters, including the enhanced gait variability index (eGVI), could be used to sensitively discriminate between different severities of DCM. A total of 153 patients recently diagnosed with DCM were recruited and stratified on the basis of DCM severity grades, as measured using the modified Japanese Orthopedic Association (mJOA) scale. Demographic information and neurological status were collected. Gait assessments were performed using an 8 m walkway. Spearman rank correlation was used to identify relationships between gait parameters and mJOA values as well as the mJOA lower extremity (LE) subscore. Kruskal–Wallis H test was performed to evaluate differences between severity groups, as defined by mJOA classification. A significant and relatively strong correlation was found between the mJOA score and eGVI, as well as between the LE subscore of the mJOA and eGVI. Significant differences in the eGVI ($X^2(2, N = 153) = 55.04, p < 0.0001, \varepsilon2 = 0.36$) were found between all groups of DCM severity, with a significant increase in the eGVI as DCM progressed from mild to moderate. The eGVI was the most discriminative gait parameter, which facilitated objective differentiation between varying severities of DCM. Quantitative gait assessments show promise as an accurate and objective tool to diagnose and classify DCM, as well as to potentially evaluate the impact of therapeutic interventions.

Keywords: degenerative cervical myelopathy; physical impairment; gait; locomotion; gait assessment; enhanced gait variability index

---

1. Introduction

Degenerative Cervical Myelopathy (DCM) is a disorder involving chronic compression of the cervical spinal cord and is the most common form of spinal cord impairment in adults [1]. DCM can result from a wide range of pathologies, including degenerative disc disease, spondylosis, and hypertrophy or ossification of the spinal ligaments [2–6]. Cervical cord compression leads to nerve damage over time, resulting in loss of function and reduced quality of life [3,4,7,8]. Patients diagnosed with DCM usually present with at least one of the following symptoms: weakness and/or numbness of the upper extremities, reduced manual dexterity, gait and balance impairment, lower extremity spasticity, neuropathic pain, and bowel/bladder dysfunction. Although DCM is common, its detection can be challenging, as impairment can be quite subtle during the mild stage of the disease.

Early diagnosis and management of DCM are important to accord appropriate care for those living with the condition. Current clinical methods for diagnosing DCM in the early stage or when the patient presents with mild symptoms are limited to subjective history taking and clinical assessment. Objective gait assessment can potentially detect early impairment. During gait, the center of mass is propelled forward as the body alternates between periods of single and double support, which produces challenges to the overall stability of the individual. While healthy adults can successfully walk with little difficulty, one of the cardinal symptoms of DCM is impaired gait [9–11]. In DCM, gait impairment is believed to be multifactorial, including upper motor neuron and proprioceptive dysfunction. The exact mechanisms have yet to be elucidated. However, the rubrospinal, reticulospinal, and vestibulospinal tracts are descending tracts that play a role in the stability of posture and gait and are likely implicated in DCM [12–14]. Gait impairment, particularly in the early stages of DCM, often presents as subtle instability in gait and balance, rather than gross and obvious impairments related to weakness or spasticity.

Clinically, DCM is classified using the modified Japanese Orthopaedic Association scale (mJOA) [15]; with the lower extremity subscore of the mJOA describing gait impairment. The parameters that define these subtle deficits in gait are quite different from the spatiotemporal parameters that typically uncover gait impairment related to stroke or musculoskeletal issues. Therefore, we aimed to characterize the gait impairment of study participants with DCM to define and detect the specific changes resulting from progressive cervical spinal cord compression. There is evidence that individuals with moderate and severe DCM demonstrate slower gait speed, prolonged double support time, and reduced cadence, as compared to individuals lacking any physical impairments [10,16]. These adaptations serve to increase stability in DCM patients and to lower the risk of falling. Current literature has focused on either kinematics and gait parameters in patients with DCM requiring surgical intervention [10,17] or on postoperative walking speed [18]. It was shown that patients with DCM receiving conservative treatment have a significantly slower walking speed over time when compared to a surgical treatment group. Also, aberrant spinal alignment, including reduced cervical lordosis, head flexion, and increased anterior pelvic tilt documented in DCM patients preoperatively, lead to altered biomechanics of the lower extremities and therefore reduced walking speed, shorter stride length and stride time, as well as increased double support time [17]. Those studies involved patients with symptoms of myelopathy requiring surgical intervention. As delayed diagnosis and treatment might lead to greater disability [19], it seems to be important to focus on the early stages of myelopathy and on diagnostic tools. To date, there is no literature available which assesses gait parameters in patients with early or mild DCM and compares them to those of patients with more advanced DCM.

The objective of this study was to assess the correlation between subjective gait impairment of patients diagnosed with DCM, measured using the mJOA score and the lower extremity subscore, with objective gait parameters. Furthermore, we wanted to characterize mild, moderate, and severe DCM, as defined by the mJOA classification system, using quantitative spatiotemporal measurements of gait.

2. Materials and Methods

2.1. Study Design

We conducted a single-center, observational, cross-sectional study involving 153 patients recently diagnosed with DCM between May 2013 and December 2017. Research ethics board approval was obtained, and all participants provided informed consent before participation. Inclusion criteria for this study were the following: (1) one or more clinical signs of DCM (corticospinal motor deficits, hand atrophy, hyperreflexia, a positive Hoffman sign, upgoing plantar reflexes, lower limb spasticity, and/or gait ataxia), (2) one or more clinical symptoms of DCM (numb hands, clumsy hands, gait impairment, bilateral hand paresthesia, L'Hermitte's phenomenon, and/or weakness), and (3) MR imaging showing flattening, indentation, or circumferential compression of the spinal cord. Patients with previous cervical spine surgery, other documented neurological disease affecting gait assessment, disability of the lower extremities, or symptomatic lumbar stenosis and a Berg Balance Scale (BBS) <40 were excluded from the study. DCM severity was determined using the modified mJOA, and DCM was classified as mild, moderate or severe [1,2]. Demographic information, neurological examination, and BBS results to assess static and dynamic instability were documented. The control group comprised 13 healthy subjects without gait disorders, matched for age and gender, with a mean age of 56.8 ± 6.8 years. Gait data acquired from the healthy controls were used to calculate baseline values for spatiotemporal gait parameters.

2.2. Scores

The mJOA consists of four categories with a maximum of 18 possible points: upper extremity motor dysfunction (5 possible points), lower extremity motor dysfunction (7 points, see Table 1), sensory impairment of the upper limbs (3 points), and bladder dysfunction (3 points). The study participants were evaluated at initial diagnosis with a score of 18, representing no functional deficit [15]. Mild DCM was defined by mJOA values between 15 and 17, moderate DCM by mJOA values from 12 to 14, and severe DCM by a mJOA score <12. [20].

Table 1. Lower extremity subscore of modified Japanese Orthopaedic Association scale (mJOA).

	0	Complete loss of movement and sensation
	1	Complete loss of movement, some sensation present
	2	Inability to walk, but some movement
	3	Able to walk on flat ground with walking aid
Lower Extremity Subscore (/7)	4	Able to walk without walking aid but must hold a handrail on stairs
	5	Moderate to severe walking imbalance, but able to perform stairs without handrail
	6	Mild imbalance when standing OR walking
	7	Normal walking

The BBS measures balance impairment through 14 items scored from 0 to 4 points each and measures explicitly unsupported standing and sitting balance, as well as transfers [21]. A BBS of 40 has been used as a cut-off for independent ambulation [21].

2.3. Gait Assessment

Gait assessment was performed in a standardized way for all participants. After careful instruction and a "warming-up" walk back and forth, patients were asked to walk across an 8-m walkway with an integrated pressure mat four times, barefoot and at a self-selected pace. Walking aids were not allowed. All gait assessments were conducted using either the GAITRite [22] (122 subjects, Franklin, NJ, USA) or the ProtoKinetics Zeno Walkway [23] (32 subjects, 13 control group subjects, Havertown, PA, USA). ProtoKinetics Movement Analysis (PKmas) software version 5.08C3i1 (Havertown, PA, USA) was used to collect gait data from both walkway systems; this software has been previously validated against the GAITRite walkway system [3]. Spatiotemporal gait parameters are presented in Table 2 and Figure 1.

Table 2. Spatiotemporal gait parameters.

Parameter	Description	Unit
Velocity	Walking speed = distance per time	cm/s
Cadence	Steps per minute	steps/min
Base of support	Step width = perpendicular distance between two points on both feet measured during two consecutive steps	m
Step length	Distance between ground contact of one foot and the next subsequent ground contact of the opposite foot in the direction of progression	m
Stride length	Distance between ground contact of one foot and the next subsequent ground contact of the same foot in the direction of progression	m
Step time	Time between ground contact of one foot and the next subsequent ground contact of the opposite foot	s
Single-stance time	Time during gait cycle while one foot is on the ground	s
Double-stance time	Time during gait cycle while two feet are on the ground	s
Total stance time	Time that passes during single and double support of the stance phase of one extremity during a gait cycle	s
eGVI	enhanced gait variability index (includes 5 spatiotemporal gait parameters: step time, step length, step velocity, total stance time, single-stance time)	

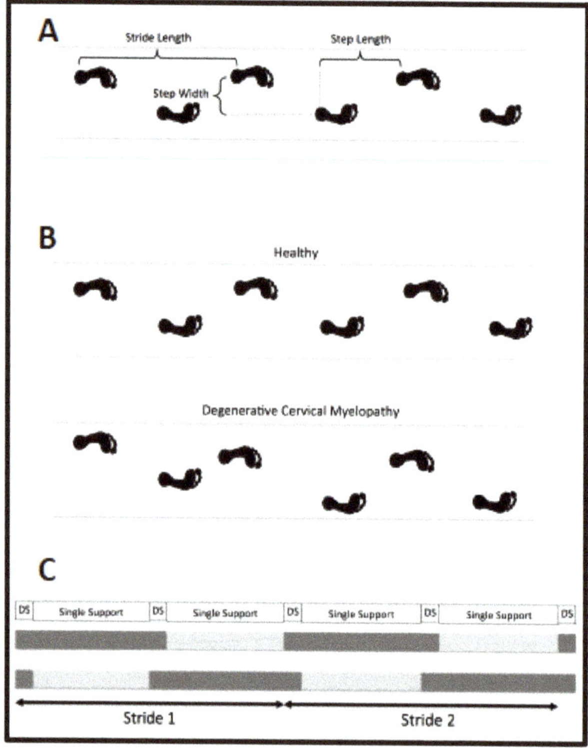

Figure 1. Visualization of spatial gait parameters (**A**). Visualization of gait variability in healthy subjects and increased variability in degenerative cervical myelopathy patients (**B**). Visualization of temporal gait parameters (**C**).

2.4. Enhanced Gait Variability Index

The enhanced gait variability index (eGVI) is an improved version of the gait variability index, including a composite of measures of gait variability based on measured spatiotemporal parameters [24]. It is used to assess the quality of gait. Gait variability is defined as the fluctuation of gait measures between steps. This measure quantifies the amount of variability observed in an individual and compares it to that of a reference group. Five spatiotemporal parameters are taken into account for the calculation of eGVI: step length, step time, stance time, single-stance time, stride velocity. The weighted variability is then transformed into a score, with 100 representing the mean gait variability, and 10 representing 1 standard deviation from the mean in a reference population [25]. The gait variability index correlates well with clinical outcomes [26]. The eGVI is an advanced version of the GVI after correction of the directional specificity and magnitude problems detected when using the GVI in assessing GV [24]. The eGVI score was calculated as an average of the left and right variability index using the ProtoKinetics Movement Analysis (PKmas) software version 5.08C3i1 (Havertown, PA, USA).

2.5. Statistics

Descriptive statistics were conducted for all parameters and are presented in mean ± SD.

Shapiro Wilk test was used to test for normality. Levene's test was used to assess the homogeneity of variance. To identify differences between DCM severity groups and acquired normative data, a one-way Kruskal–Wallis H test was conducted. A post-hoc test with Bonferroni correction was performed in the case of significance. Epsilon square was used as an effect size to indicate the magnitude of the difference between the severity groups. Spearman's rank correlation coefficient was used to identify relationships between quantitative gait parameters and both the mJOA values as well as the mJOA lower extremity subscore (see Figure 2). The significance level was set at $p \leq 0.05$. Statistical analysis was performed using R Version 3.6.1.

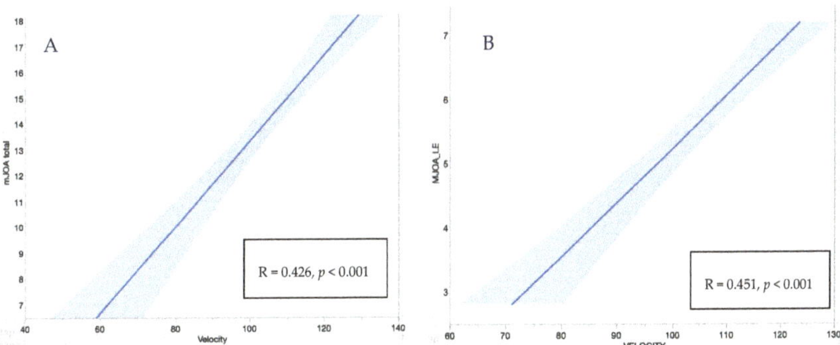

Figure 2. We observed a significant decrease in velocity with decreased mJOA score (**A**) and decreased mJOA LE subscore (**B**).

3. Results

3.1. Patient Demographics

The sample of DCM patients consisted of 83 male and 70 female participants, with a mean age of 56.81 ± 10.92 years. The mean duration of symptoms was 44.19 ± 56.06 months prior to assessment. Table 3 defines the sample stratified by mJOA into mild, moderate, and severe groups, also reporting the mean values and standard deviations of spatiotemporal gait parameters and eGVI. We found that 48.7% of patients in the mild DCM group, 21.2% in the moderate DCM group, and 0% in the severe DCM group presented within the range of eGVI of our control group. In addition, 35.9% of patients with mild DCM, 5.0% with moderate DCM, and 0% with severe DCM presented step length within the range of the control group.

Table 3. Mean (± SD) of patient and gait specific parameters, stratified by the modified Japanese Orthopaedic Association (mJOA) scale. mJOA LE: mJOA lower extremity.

Variable	Control Group n = 13	Mild DCM n = 82	Moderate DCM n = 40	Severe DCM n = 31	All DCM n = 153
Age	56.75 (6.77)	55.3 (11.01)	55.73 (9.75)	62.19 (10.91)	56.81 (10.92)
mJOA Score		15.92 (0.73)	13.13 (0.82)	9.94(2.5)	13.98 (2.50)
mJOA LE Subscore		6.51 (0.55)	5.10 (1.12)	3.71 (1.35)	5.58 (1.35)
Berg Balance Score		53.52 (5.24)	49.63 (7.09)	42.59 (4.65)	47(6.1)
Velocity (cm/sec)	119.22 (11.61)	114.84 (23.71)	106.44 (23.72)	74.18 (29.51)	104.41 (29.51)
Cadence (steps/min)	114.74(9.49)	111.49 (12.67)	108.58 (12.46)	92.55(15.98)	106.89 (15.99)
Base of Support (cm)	8.16 (3.74)	9.13 (3.28)	8.21 (3.85)	9.24 (3.65)	8.91 (3.65)
Step Length (cm)	63.57 (4.87)	60.81 (9.66)	57.94 (8.83)	45.96 (11.44)	57.05 (11.44)
Total Stance Time (sec)	0.649 (0.12)	0.702 (0.10)	0.717 (0.09)	0.905 (0.25)	0.747 (0.163)
Single-Support Time (sec)	0.410 (0.03)	0.389 (0.04)	0.395 (0.04)	0.409 (0.07)	0.394 (0.046)
Double-Support Time (sec)	0.249 (0.029)	0.303 (0.08)	0.316 (0.06)	0.485 (0.22)	0.343 (0.138)
Single-Stance Ratio	1.56 (0.20)	1.35 (0.29)	1.29 (0.23)	0.99 (0.38)	1.26 (0.32)
Enhanced Gait Variability Index	103.36(4.54)	110.9 (9.73)	119.14 (10.14)	132.94 (12.78)	117.54 (13.5)

3.2. Quantitative Assessment of Gait Parameters

Table 4 shows the correlation between gait parameters and mJOA values as well as mJOA lower extremity subscores. A significant relatively strong correlation was found between the subjective mJOA lower extremity subscore and eGVI ($|R| = 0.567, p < 0.05$) as well as velocity ($|R| = 0.456, p < 0.05$). Also, a significant relatively strong correlation was found between mJOA score and eGVI ($|R| = 0.551, p < 0.05$). A significant but moderate correlation was found between mJOA score and velocity ($|R| = 0.426, p < 0.05$), as shown in Table 4 and Figure 2.

Table 4. Spearman's rank correlation coefficients between gait parameters compared with mJOA LE subscore and mJOA score (total). Confidence interval was set to 95%.

Gait Parameters	mJOA LE	*p*−Value	mJOA	*p*−Value
Velocity (cm/sec)	0.456	<0.001	0.426	<0.001
Cadence (steps/min)	0.346	<0.001	0.286	<0.001
Base of Support (cm)	0.044	0.6	0.038	0.6
Step Length (cm)	0.434	<0.001	0.417	<0.001
Total Stance Time (sec)	−0.352	<0.001	−0.303	<0.001
Single-Support Time (sec)	−0.058	0.47	0.004	0.959
Double-Support Time (sec)	−0.404	<0.001	−0.382	<0.001
Single-Stance Ratio	0.413	<0.001	0.417	<0.001
Enhanced Gait Variability Index	−0.567	<0.001	−0.551	<0.001

The Kruskal–Wallis test showed a significant difference in gait variability ($X^2(2, N = 153) = 55.04$, $p < 0.0001$, $\varepsilon^2 = 0.36$). A post-hoc test using Dunn's test with Bonferroni correction showed a significant increase in variability for more severe stages of DCM ($p < 0.001$) and a strong effect size ($\varepsilon^2 = 0.36$). We found a mean score of 111.18 ± 9.85 for mild DCM versus a mean score of 119.14 ± 10.14 for moderate DCM (mild/moderate = $p < 0.001$) and a mean of 132.94 ± 12.78 for severe DCM (moderate/severe = $p < 0.001$). We also detected a significant difference in velocity ($X^2(2, N = 153) = 35.59, p < 0.0001, \varepsilon^2 = 0.23$), stride velocity ($X^2(2, N = 153) = 32.79, p < 0.0001, \varepsilon^2 = 0.22$), and step length ($X^2(2, N = 153) =$

30.23, $p < 0.0001$, $\varepsilon^2 = 0.19$) between patients with moderate and severe DCM, as shown in Table 5 and Figure 3.

Table 5. Kruskal–Wallis H-test, Bonferroni-adjusted p, and Epsilon squared effect sizes.

Gait Parameter	H(df)	p	Padj Mild/Moderate	Padj Mild/Severe	Padj Moderate/Severe	Epsilon^2
Velocity	35.59(2)	<0.0001	0.081	<0.0001	0.001	0.23 [+]
Cadence	22.92(2)	<0.0001	0.59	<0.0001	0.004	0.15
Base of support	2.73(2)	0.26	-	-	-	0.02
Step Length	30.23(2)	<0.0001	0.25	<0.0001	0.002	0.19 [+]
Stride Velocity	32.79(2)	<0.0001	0.08	<0.0001	0.003	0.22 [+]
Total Stance Time	21.80(2)	0.0002	0.72	<0.0001	0.005	0.14
Single-Support Time	1.83(2)	0.4	-	-	-	0.01
Double-Support Time	25.54(2)	<0.0001	0.34	<0.0001	0.0043	0.16
Single-Stance Ratio	25.96(2)	<0.0001	0.59	<0.0001	0.002	0.17
eGVI	55.04(2)	<0.0001	0.001*	<0.0001 *	0.001 *	0.36 [++]
Age	9.22(2)	0.01	1	0.012	0.023	0.06

* significant difference, [++] strong effect size, [+] relatively strong effect size.

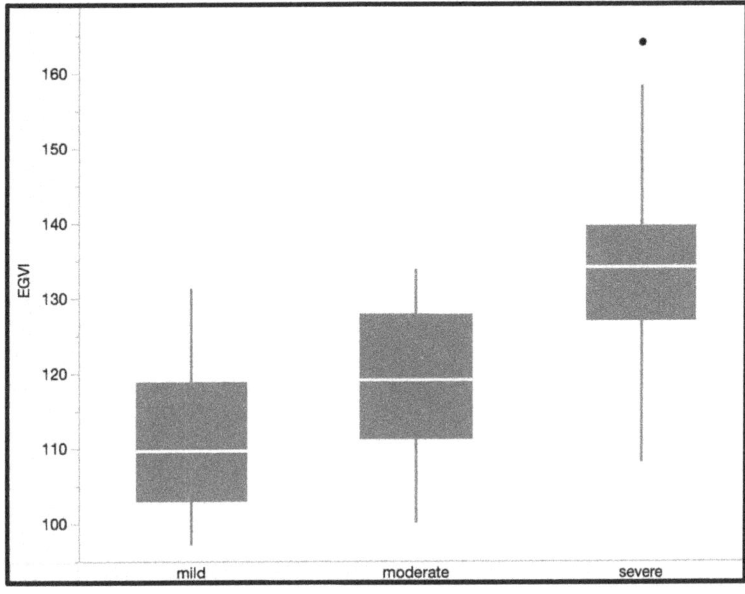

Figure 3. A significant increase in gait variability as measured by the eGVI was observed between severity groups in degenerative cervical myelopathy (DCM) patients.

4. Discussion

To our knowledge, this is the largest study to date characterizing specific differences in gait parameters between severity subtypes of DCM. Additionally, this study assessed the correlation between objective spatiotemporal gait parameters and subjective clinical gait impairment in patients with DCM.

We found significant differences between control subjects and patients with mild, moderate, and severe DCM. Specifically, the enhanced gait variability index proved to be a useful tool to document significant differences between all severity groups as defined by the mJOA. Mean eGVI increased significantly from 103.36 ± 4.54 in the control group to 110.9 ± 9.73 in patients with mild DCM, 119.14 ± 10.14 in patients with moderate DCM, and 132.94 ± 12.78 in patients with severe DCM. Based on the literature [24], an eGVI of approximately 100 is within normative range, and a clinically relevant difference occurs when there is a change of at least 10 points (one SD).

Gait deficits are commonly self-reported and, at times, objectively measured with timed walking tests [18]. The primary screening tool used to evaluate individuals with DCM is the mJOA scale [15], which is a subjective clinical score that also assesses walking difficulties. The mJOA score is used to stratify DCM by severity, with mild DCM represented by a score of 15–17, moderate DCM by a score of 12–14, and severe DCM by a score <12 [20]. The lower extremity subscore is presented in Table 3 and Figure 2. Patients with mild DCM commonly report only minimal gait impairment, and in these cases, gait deficits are typically not detectable with routine clinical exams [20,27]. Timed walking tests can detect changes in gait speed; however, for individuals with mild DCM, gait velocity typically falls within a normative range, meaning that subtle impairments cannot be quantified. While the subtle deficits do not have a definitive impact on function, identifying these changes can be essential for early identification of the disease and monitoring disease progression.

In contrast to the large amount of literature surrounding gait analysis in other neurological conditions [28], such as stroke [29], Parkinson's disease, and other neurological conditions [26], little is known about specific spatiotemporal gait parameters in DCM [9,28]. In a recent publication, Zheng et al. [30] evaluated the correlation between the JOA score and specific gait parameters in patients with DCM and lumbar disc herniation (LDH). They found only a weak correlation between the JOA score and step duration, cycle duration, double-support time, gait speed, cadence, and stride length and no correlation with single-support time. In a multiple regression analysis, they only found the lower extremity motor function subscore as a significant but weakly correlated parameter, but no significant factor was associated with the motor function of lower extremities. In contrast, our study shows a significant and moderately strong correlation between the mJOA score and both velocity and step length, as well as a relatively strong correlation between the mJOA score and the eGVI. Zheng et al. [30] state that the JOA scoring system might not adequately reflect gait impairment and that gait analysis might be more reliable in detecting walking impairment. Since they found a better correlation between the JOA lumbar score and gait parameters, they suggested that the difference might be due to the use of fewer questions regarding walking in the JOA cervical score. In contrast, we were able to demonstrate a significant correlation between the mJOA score and various spatiotemporal gait parameters. This might be due to the use of the mJOA scale in our study, where the emphasis on walking was improved.

We also found significantly reduced velocity, stride velocity, and step length between moderate and severe DCM groups. Singh et al. found a continuous decrease in walking speed with time after the initial diagnosis of DCM and a significantly increased walking speed after cervical decompressive surgery [18] using a 30-m walking test. In comparison, Haddas et al. [17] found significantly decreased cadence, velocity, single-support time, step length, and step width in patients with DCM compared to a healthy control group. They only assessed patients already scheduled for decompression surgery, with more severe DCM, which explains the higher correlation in most gait parameters. Decreased velocity, step velocity, and step length, are most likely related to decreased balance while walking as DCM progresses. Reduced velocity and step length will help increase gait stability and, therefore, also decrease gait variability. This might cover part of the variability in gait parameters, especially at mild stages of DCM, and explains the different results in comparison to those of Haddas et al.

This study has several limitations. Patients in the severe DCM group were significantly older than patients in the other two groups (mean age of 62 in severe DCM versus a mean age of 55 in mild and moderate DCM). This age difference might contribute to further changes in gait assessment in comparison to younger subjects. Virmani et al. [31] were able to show a significantly increased variability in stride length with age during steady-state gait using stepwise multiple regression analysis. When using univariate analysis, an increase in stride velocity was also detected. Velocity and stride velocity also significantly decreased with more severe DCM in our dataset. This was observed not only when comparing patients with moderate and severe DCM, but also when comparing the control group with patients with mild and moderate DCM, although the last three groups did not present a significant age difference. Another drawback is that we used average data between the left and the right leg, which might hide relevant information, especially in relation to variability assessments, but at the advantage of eliminating lower extremity-related gait patterns.

Further research is necessary to evaluate and compare pre- and postoperative/post-treatment gait parameters in patients diagnosed with DCM. This can provide further insight into subtle changes associated with disease progression and treatments. The authors believe this work is only the initial step in defining a sensitive assessment that can characterize gait impairment in DCM patients. We look to continuing developing these findings into more validated and psychometrically sound parameters as the measures continue to be used and implemented in clinical/research environments.

Author Contributions: Conceptualization, S.K.-R. and M.G.F.; Methodology, S.K.-R., A.C.R., A.R.M., M.A.A., and J.H.B.; Software, A.C.R., M.A.A., and A.R.M.; Validation, S.K.-R., A.C.R., M.A.A., A.R.M., E.M.M., and M.G.F.; Formal Analysis, A.C.R., M.A.A., and A.R.M.; Investigation, S.K.-R. and M.G.F.; Resources, S.K.-R. and M.G.F.; Data Curation, S.K.-R., A.C.R., A.R.M., M.A.A., A.R.M., J.H.B., E.M.M., L.R., C.C., and D.J.; Writing—Original Draft Preparation, S.K.-R. and A.C.R.; Writing—Review & Editing, S.K.-R., A.C.R., and M.G.F.; Visualization, A.C.R.; Supervision, S.K.-R. and M.G.F.; Project Administration, S.K.-R. and M.G.F.; Funding Acquisition, S.K.-R. and M.G.F. All authors have read and agreed to the published version of the manuscript.

Funding: This research work was supported by the Toronto Western Hospital Spine Program, AOspine North America, and the Cervical Spine Research Society.

Acknowledgments: This work was supported by the Toronto Western Hospital Spine Program, AOspine North America, and the Cervical Spine Research Society. We thank Tim Worden for editing and visualization of gait parameters. Furthermore, we thank our study participants for volunteering to be a part of this project.

Conflicts of Interest: The authors declare no conflict of interest.

References

1. Kalsi-Ryan, S.; Karadimas, S.K.; Fehlings, M.G. Cervical spondylotic myelopathy: The clinical phenomenon and the current pathobiology of an increasingly prevalent and devastating disorder. *Neuroscientist* **2013**, *19*, 409–421. [CrossRef] [PubMed]
2. Fehlings, M.G.; Tetreault, L.; Hsieh, P.C.; Traynelis, V.; Wang, M.Y. Introduction: Degenerative cervical myelopathy: Diagnostic, assessment, and management strategies, surgical complications, and outcome prediction. *Neurosurg. Focus* **2016**, *40*, E1. [CrossRef] [PubMed]
3. Nouri, A.; Tetreault, L.; Singh, A.; Karadimas, S.K.; Fehlings, M.G. Degenerative Cervical Myelopathy: Epidemiology, Genetics, and Pathogenesis. *Spine* **2015**, *40*, E675–E693. [CrossRef] [PubMed]
4. Ray, S.K.; Matzelle, D.D.; Wilford, G.G.; Hogan, E.L.; Banik, N.L. Increased calpain expression is associated with apoptosis in rat spinal cord injury: Calpain inhibitor provides neuroprotection. *Neurochem. Res.* **2000**, *25*, 1191–1198. [CrossRef]
5. Tetreault, L.; Goldstein, C.L.; Arnold, P.; Harrop, J.; Hilibrand, A.; Nouri, A.; Fehlings, M.G. Degenerative Cervical Myelopathy: A Spectrum of Related Disorders Affecting the Aging Spine. *Neurosurgery* **2015**, *77* (Suppl. 4), S51–S67. [CrossRef]
6. Young, W.F. Cervical spondylotic myelopathy: A common cause of spinal cord dysfunction in older persons. *Am. Fam. Physician* **2000**, *62*, 1064–1070, 1073.
7. King, J.T., Jr.; McGinnis, K.A.; Roberts, M.S. Quality of life assessment with the medical outcomes study short form-36 among patients with cervical spondylotic myelopathy. *Neurosurgery* **2003**, *52*, 113–120; discussion 121.
8. Tracy, J.A.; Bartleson, J.D. Cervical spondylotic myelopathy. *Neurologist* **2010**, *16*, 176–187. [CrossRef]
9. Malone, A.; Meldrum, D.; Bolger, C. Gait impairment in cervical spondylotic myelopathy: Comparison with age- and gender-matched healthy controls. *Eur. Spine J.* **2012**, *21*, 2456–2466. [CrossRef]

10. Kuhtz-Buschbeck, J.P.; Johnk, K.; Mader, S.; Stolze, H.; Mehdorn, M. Analysis of gait in cervical myelopathy. *Gait Posture* **1999**, *9*, 184–189. [CrossRef]

11. Siasios, I.D.; Spanos, S.L.; Kanellopoulos, A.K.; Fotiadou, A.; Pollina, J.; Schneider, D.; Becker, A.; Dimopoulos, V.G.; Fountas, K.N. The Role of Gait Analysis in the Evaluation of Patients with Cervical Myelopathy: A Literature Review Study. *World Neurosurg.* **2017**, *101*, 275–282. [CrossRef]

12. Kidd, G.; Lawes, N.; Musa, I. *Understanding Neuromuscular Plasticity: A Basic for Clinical Rehabilitation*; Edward Arnolds: London, UK, 1992.

13. Drew, T.; Prentice, S.; Schepens, B. Cortical and brainstem control of locomotion. *Prog. Brain Res.* **2004**, *143*, 251–261.

14. Dietz, V. Human neuronal control of automatic functional movements: Interaction between central programs and afferent input. *Physiol. Rev.* **1992**, *72*, 33–69. [CrossRef] [PubMed]

15. Kopjar, B.; Fehlings, M.; Hanson, B. Validity of the Modified Japanese Orthopedic Association Score in Patients with Cervical Spondylotic Myelopathy: The AOSpine North America Multicenter Prospective Study. *Spine J.* **2011**, *11*, S73–S74. [CrossRef]

16. Kim, C.R.; Yoo, J.Y.; Lee, S.H.; Lee, D.H.; Rhim, S.C. Gait analysis for evaluating the relationship between increased signal intensity on t2-weighted magnetic resonance imaging and gait function in cervical spondylotic myelopathy. *Arch. Phys. Med. Rehabil.* **2010**, *91*, 1587–1592. [CrossRef] [PubMed]

17. Haddas, R.; Patel, S.; Arakal, R.; Boah, A.; Belanger, T.; Ju, K.L. Spine and lower extremity kinematics during gait in patients with cervical spondylotic myelopathy. *Spine J.* **2018**, *18*, 1645–1652. [CrossRef]

18. Singh, A.; Choi, D.; Crockard, A. Use of walking data in assessing operative results for cervical spondylotic myelopathy: Long-term follow-up and comparison with controls. *Spine* **2009**, *34*, 1296–1300. [CrossRef]

19. Pope, D.H.; Mowforth, O.D.; Davies, B.M.; Kotter, M.R.N. Diagnostic Delays Lead To Greater Disability In Degenerative Cervical Myelopathy and Represent A Health-Inequality. *Spine* **2020**, *45*, 368–377. [CrossRef]

20. Tetreault, L.; Kopjar, B.; Nouri, A.; Arnold, P.; Barbagallo, G.; Bartels, R.; Qiang, Z.; Singh, A.; Zileli, M.; Vaccaro, A.; et al. The modified Japanese Orthopaedic Association scale: Establishing criteria for mild, moderate and severe impairment in patients with degenerative cervical myelopathy. *Eur. Spine J.* **2017**, *26*, 78–84. [CrossRef]

21. Berg, K.O.; Wood-Dauphinee, S.L.; Williams, J.I.; Maki, B. Measuring balance in the elderly: Validation of an instrument. *Can. J. Public Health = Rev. Can. Sante Publique* **1992**, *83* (Suppl. 2), S7–S11.

22. Menz, H.B.; Latt, M.D.; Tiedemann, A.; Mun San Kwan, M.; Lord, S.R. Reliability of the GAITRite walkway system for the quantification of temporo-spatial parameters of gait in young and older people. *Gait Posture* **2004**, *20*, 20–25. [CrossRef]

23. Vallabhajosula, S.; Humphrey, S.K.; Cook, A.J.; Freund, J.E. Concurrent Validity of the Zeno Walkway for Measuring Spatiotemporal Gait Parameters in Older Adults. *J. Geriatr. Phys. Ther. (2001).* **2019**, *42*, E42–E50. [CrossRef] [PubMed]

24. Gouelle, A.; Rennie, L.; Clark, D.J.; Megrot, F.; Balasubramanian, C.K. Addressing limitations of the Gait Variability Index to enhance its applicability: The enhanced GVI (EGVI). *PLoS ONE* **2018**, *13*, e0198267. [CrossRef] [PubMed]

25. Gouelle, A.; Megrot, F.; Presedo, A.; Husson, I.; Yelnik, A.; Pennecot, G.F. The gait variability index: A new way to quantify fluctuation magnitude of spatiotemporal parameters during gait. *Gait Posture* **2013**, *38*, 461–465. [CrossRef]

26. Moon, Y.; Sung, J.; An, R.; Hernandez, M.E.; Sosnoff, J.J. Gait variability in people with neurological disorders: A systematic review and meta-analysis. *Hum. Mov. Sci.* **2016**, *47*, 197–208. [CrossRef]

27. Fehlings, M.G.; Wilson, J.R.; Kopjar, B.; Yoon, S.T.; Arnold, P.M.; Massicotte, E.M.; Vaccaro, A.R.; Brodke, D.S.; Shaffrey, C.I.; Smith, J.S.; et al. Efficacy and safety of surgical decompression in patients with cervical spondylotic myelopathy: Results of the AOSpine North America prospective multi-center study. *J. Bone Jt. Surg. Am. Vol.* **2013**, *95*, 1651–1658. [CrossRef]

28. Chen, G.; Patten, C.; Kothari, D.H.; Zajac, F.E. Gait differences between individuals with post-stroke hemiparesis and non-disabled controls at matched speeds. *Gait Posture* **2005**, *22*, 51–56. [CrossRef]

29. Guzik, A.; Druzbicki, M.; Przysada, G.; Wolan-Nieroda, A.; Szczepanik, M.; Bazarnik-Mucha, K.; Kwolek, A. Validity of the gait variability index for individuals after a stroke in a chronic stage of recovery. *Gait Posture* **2019**, *68*, 63–67. [CrossRef]

30. Zheng, C.F.; Liu, Y.C.; Hu, Y.C.; Xia, Q.; Miao, J.; Zhang, J.D.; Zhang, K. Correlations of Japanese Orthopaedic Association Scoring Systems with Gait Parameters in Patients with Degenerative Spinal Diseases. *Orthop. Surg.* **2016**, *8*, 447–453. [CrossRef]
31. Virmani, T.; Gupta, H.; Shah, J.; Larson-Prior, L. Objective measures of gait and balance in healthy non-falling adults as a function of age. *Gait Posture* **2018**, *65*, 100–105. [CrossRef]

Journal of
Clinical Medicine

Article

The Relationship Between Gastrointestinal Comorbidities, Clinical Presentation and Surgical Outcome in Patients with DCM: Analysis of a Global Cohort

Aria Nouri [1,2], Jetan H. Badhiwala [3], So Kato [4], Hamed Reihani-Kermani [5], Kishan Patel [6], Jefferson R. Wilson [3], Insa Janssen [1], Joseph S. Cheng [2], Karl Schaller [1], Enrico Tessitore [1] and Michael G. Fehlings [3,*]

[1]   Department of Neurosurgery, University of Geneva, 1205 Geneva, Switzerland;
       arianouri9@gmail.com (A.N.); InsaKatrin.Janssen@hcuge.ch (I.J.); Karl.Schaller@hcuge.ch (K.S.);
       enrico.tessitore@hcuge.ch (E.T.)
[2]   Department of Neurosurgery, University of Cincinnati College of Medicine, Cincinnati, OH 45267-0515,
       USA; chengj6@ucmail.uc.edu
[3]   Division of Neurosurgery, University of Toronto, Toronto, ON M5S 1A1, Canada;
       jetan.badhiwala@gmail.com (J.-H.B.); jeffersonwilson7@gmail.com (J.R.W.)
[4]   Department of Orthopaedic Surgery, University of Tokyo, Tokyo 100068, Japan; sokato34@gmail.com
[5]   Neuroscience Research Center, Kerman University of Medical Sciences, Kerman, Iran;
       h_reihani@hotmail.com
[6]   Department of Neurosurgery, Yale University, New Haven, CT 06520-8082, USA; kishan.patel@yale.edu
*   Correspondence: Michael.Fehlings@uhn.on.ca; Tel.: 416-603-5627; Fax: 416-603-5298

Received: 31 December 2019; Accepted: 18 February 2020; Published: 26 February 2020

Abstract: Degenerative cervical myelopathy (DCM) is the most common cause of spinal cord impairment in adults, presenting most frequently in patients 50 years or older. Gastrointestinal comorbidities (GICs) commonly occur in this group; however, their relationship with DCM has not been thoroughly investigated. It is the objective of the present study to investigate the difference between patients with or without GICs who are surgically treated for DCM. A cohort of 757 patients with clinical data and 458 with magnetic resonance imaging (MRI) data from the AOSpine North America and AOSpine International studies on DCM was evaluated. GICs were obtained at presentation and included gastric, intestinal, hepatic, and pancreatic conditions. Patients were dichotomized into 2 groups: those with GICs and those without GICs. Both clinical and MRI presentation, as well as baseline neurological and functional status, were compared. Neurological and functional outcomes at 2-year follow-up were also compared. GICs were present in 121 patients (16%). These patients were less commonly male (48.76% vs. 65.4%, $p = 0.001$) and were slightly less neurologically impaired based on the Nurick grade (3.05 ± 1.10 vs. 3.28 ± 1.16, $p = 0.044$) but not based on mJOA (12.74 ± 2.62 vs. 12.48 ± 2.76, $p = 0.33$). They also had a worse physical health score (32.80 ± 8.79 vs. 34.65 ± 9.38 $p = 0.049$), worse neck disability (46.31 ± 20.04 vs. 38.23 ± 20.44, $p < 0.001$), a lower prevalence of upper motor neuron signs (hyperreflexia, 70.2% vs. 78.9%, $p = 0.037$; Babinski's sign 24.8% vs. 37.3%, $p = 0.008$), and a higher rate of psychiatric comorbidities (31.4% vs. 10.4%, $p < 0.0001$). On MRI, GIC patients less commonly exhibited signal intensity changes (T2 hyperintensity, 49.2% vs. 75.6%, $p < 0.001$; T1 hypointensity, 9.7% vs. 21.1%, $p = 0.036$), and had a lower number of T2 hyperintensity levels (0.82 ± 0.98 vs. 1.3 ± 1.11, $p = 0.001$). There was no difference in surgical outcome between the groups. DCM patients with GICs are more likely to be female and have significantly more general health impairment and neck disability. However, these patients have less clinical and MRI features typical of more severe neurological impairment. This constellation of symptoms is considerably different than those typically observed in DCM, and it is therefore plausible that nutritional factors may contribute to this unique observation.

Keywords: cervical spondylotic myelopathy (CSM); prospective; multicenter; anterior; posterior

---

1. Introduction

Degenerative cervical myelopathy is the most common cause of spinal cord impairment in industrialized countries and can lead to significant neurological and functional dysfunction, as well as reduced quality of life [1]. The underlying pathology is heterogeneous and can include intervertebral disc disease, arthritic changes, hypertrophy and/or ossification of the spinal canal ligaments, and spondylolisthesis, ultimately leading to spinal cord injury through static and dynamic injury mechanisms [1]. Depending on the number of cervical levels involved, the degree of cord compression, and the natural history, patients present with a wide-ranging spectrum of clinical manifestations [2,3]. Symptoms include hyperreflexia, weakness, numbness, and loss of proprioception/balance, and clinical signs, such as Hoffmann's sign, Babinski reflex, Lhermitte's phenomenon, ankle clonus, inverted brachioradialis reflex, and Romberg's sign, which may be elicited on clinical examination [2,4,5]. Neurophysiological examination may indicate changes in motor and sensory evoked potentials; MRI signal intensity changes on T2 and T1 may highlight injury to the spinal cord. These various clinical factors and examinations have been used to assess degree of neurological impairment and surgical outcome. However, relatively little research has been undertaken to assess how comorbidities, such as gastrointestinal disease, impact baseline neurological status, and recovery potential in patients undergoing surgical treatment.

Gastrointestinal comorbidities (GICs) have the potential to influence the presentation and recovery of patients with myelopathy in a number of ways. For example, GICs can result in malnutrition (such as hypocupremia), anemia through blood loss, and vitamin B12 (B12) deficiency—all of which may impact spinal cord function or surgical recovery [6,7]. With regard to B12 deficiency, it has been recently suggested that vitamin B12 (B12) deficiency may be a common and under-recognized comorbidity in patients with DCM [8], and is also a differential diagnosis [9,10]. It has also been shown that anemia is related to higher surgical morbidity, worse neurological status at baseline and neurological outcomes, higher rates of medical complications, and raises the risk of complications by increasing the probability that a patient will require an allogeneic RBC infusion [11–13]. Other studies have shown that malnutrition increases 90-day major medical complications, 1-year mortality, and is a predictor of increased infection and wound dehiscence rates after lumbar spine surgery [14].

Given that identification of sequalae of GICs may have an important impact on the clinical management of DCM patients, it is the objective of the present paper to evaluate the influence of GICs on baseline neurological function and surgical outcomes for treatment of DCM.

2. Methods

2.1. Study Data

The combined AOSpine study cohort comprises 757 patients (AOSpine North American study, $n = 278$; AOSpine International Study, $n = 479$) [15,16]. The North American study was conducted between 2005 and 2007 and included 12 North American sites (11 USA, 1 Canada); the International study was conducted between 2007 and 2011 and included 16 global sites comprising 4 regions (North America, Latin America, Asia, and Europe). The primary study objective was to assess the safety and efficacy of surgical treatment for DCM and was previously reported [15,16]. Adult patients ($\geq$18 years of age) were included if they had clinical signs and symptoms of myelopathy that were confirmed via imaging. Patients were excluded if they had an active infection, neoplastic disease, rheumatoid arthritis, ankylosing spondylitis, previous surgery, or concomitant signs of lumbar stenosis. Patient clinical data, general health (SF-36) [17], Neck Disability Index (NDI) [18], and neurological function (modified Japanese Association score [mJOA] [19] and Nurick grade [20] were assessed. The pain subscore of NDI, which ranges from 0 to 5, was assessed to specifically evaluate pain. GICs were recorded non-specifically as present or absent and included potential gastric, hepatic, pancreatic, and intestinal comorbidities. Research ethics board approval was given at each participating center, and external monitors were used to visit the sites.

2.2. MRI Data

MRI (1.5T or 3T) acquisitions were performed according to local protocols (no standardized protocols were used), and typically included axial and sagittal T2-weighted and sagittal T1-weighted images. DICOM (Digital Imaging and Communications in Medicine) and conventional image formats (JPEG, TIFF) were reviewed. DICOMs were reviewed using Osirix (www.osirix-viewer.com; Pixmeo, Geneva, Switzerland). MRIs were available for 458 patients, and the prevalence and spectrum of DCM pathology were previously published [21]. MRIs were assessed for the presence and absence of specific pathologies (e.g., isolated disc pathology, spondylolisthesis), for the presence of T2 signal hyperintensity, and T1 signal hypointensity changes. Signal intensity changes on T2 and T1 were reviewed by 3 raters, and the relationship between these changes and clinical presentation, as well as surgical outcome, were previously reported [2]. Therein, inter-rater reliability for signal changes was reported as being in substantial agreement for T2 hyperintensity (Fleiss Kappa: 0.60), and in fair agreement for T1 hypointensity (Fleiss Kappa: 0.31).

2.3. Statistical Analysis

Statistical analysis was performed with SPSS (version 25.0, IBM, Armonk, NY, USA). Patients with DCM were separated into groups comprising those with or without GICs. Continuous variables are presented as means and were compared using independent t-tests. Categorical variables are presented as proportions and were assessed using Chi square. A last observation carry-forward approach was used to impute missing data for follow-up at 2 years. Measures of neurological and functional impairment between patients with and without GICs were compared at baseline and 2-year follow-up (mean difference from baseline) using independent t-tests. The baseline pain subscore of NDI was compared using an independent t-test. As a sensitivity analysis, between-group comparisons of change in mJOA, Nurick grade, NDI, and SF-36 physical component summary (PCS) and mental component summery (MCS) from baseline were made with the use of mixed-effects models for repeated measures. Fixed effects for the presence of GICs (GICs vs. no GICs), time (1 year, 2 year), and time x GIC interaction were included. Comparisons of least-squares means between groups at each time point were performed using the appropriate contrasts within the mixed-effect models.

3. Results

There were 121 patients (16%) with GICs and 636 patients (84%) without GICs (Table 1). GIC patients were less commonly male (48.76% vs. 65.4%, $p = 0.001$) and were on average 2 years older than patients without GICs (57.98 ± 10.21 vs. 56.04 ± 12.10, $p = 0.065$); however, this did not reach statistical significance. Neurologically, GIC patients were marginally less impaired than patients without GICs (Nurick grade, 3.05 ± 1.10 vs. 3.28 ± 1.16, $p = 0.044$; mJOA, 12.74 ± 2.62 vs. 12.48 ± 2.76, $p = 0.33$) but had a higher rate of psychiatric comorbidities (31.4% vs. 10.4%, $p < 0.0001$). Patients with GICs also had worse physical disability (SF-36 PCS, 32.80 ± 8.79 vs. 34.65 ± 9.38, $p = 0.049$) and worse neck disability (NDI, 46.31 ± 20.04 vs. 38.23 ± 20.44, $p < 0.001$), but a lower prevalence of upper motor signs (hyperreflexia, 70.2% vs. 78.9%, $p = 0.037$; Babinski's sign 24.8% vs. 37.3%, $p = 0.008$). Duration of symptoms was similar for patients with and without GICs. The baseline NDI pain subscore was significantly worse in patients with GICs than those without (2.27 ± 1.32 vs. 1.75 ± 1.31, $p < 0.001$).

Table 1. Patient demographics and clinical and MRI presentation.

Clinical	Gastrointestinal Comorbidity	Non-Gastrointestinal Comorbidity	*p*-Value
Age	57.98 ± 10.21 ($n = 121$)	56.04 ± 12.10 ($n = 636$)	0.065
Sex (Male)	48.76% ($n = 59/121$)	65.4% ($n = 220/636$)	**0.001**
Duration of Symptoms	26.24 ± 38.22 ($n = 121$)	26.64 ± 39.14 ($n = 636$)	0.918
Psychiatric Comorbidities	31.4% ($n = 38/121$)	10.4% ($n = 66/636$)	**<0.0001**
mJOA	12.74 ± 2.62 ($n = 120$)	12.48 ± 2.76 ($n = 623$)	0.33
Nurick	3.05 ± 1.10 ($n = 121$)	3.28 ± 1.16 ($n = 636$)	**0.044**
NDI	46.31 ± 20.04 ($n = 100$)	38.23 ± 20.44 ($n = 560$)	**0.0003**
SF-36 PCS	32.80 ± 8.79 ($n = 116$)	34.65 ± 9.38 ($n = 618$)	**0.049**
SF-36 MCS	38.47 ± 14.01 ($n = 116$)	40.57 ± 13.43 ($n = 618$)	0.124
Clinical Signs and Symptoms ($n = 756$)			
Numb Hands	92.6% ($n = 112/121$)	88.2% ($n = 560/635$)	0.16
Clumsy Hands	71.9% ($n = 87/121$)	74.6% ($n = 474/635$)	0.53
Impairment of Gait	74.4% ($n = 90/121$)	75.4% ($n = 479/635$)	0.81
Bilateral arm paresthesia	61.2% ($n = 74/121$)	55.7% ($n = 354/635$)	0.27
Lhermitte's Phenomena	29.8% ($n = 36/121$)	26.0% ($n = 165/635$)	0.39
Weakness	83.5% ($n = 101/121$)	82.2% ($n = 522/635$)	0.74
Corticospinal motor deficits	62.0% ($n = 75/121$)	62.5% ($n = 397/635$)	0.91
Atrophy of hand intrinsic muscles	35.5% ($n = 43/121$)	35.9% ($n = 228/635$)	0.94
Hyperreflexia	70.2% ($n = 85/121$)	78.9% ($n = 501/635$)	**0.037**
Hoffmann's Sign	58.7% ($n = 71/121$)	62.7% ($n = 398/635$)	0.41
Babinski's sign	24.8% ($n = 30/121$)	37.3% ($n = 237/635$)	**0.008**
Lower limb spasticity	40.5% ($n = 49/121$)	47.9% ($n = 304/635$)	0.14
Broad-based, unstable gait	55.4% ($n = 67/121$)	59.1% ($n = 375/635$)	0.45
MRI			
Anterior–Posterior Compression	53.7% ($n = 36/67$)	59.8% ($n = 234/391$)	0.35
Number of Levels Compressed	2.91 ± 1.25 ($n = 67$)	3.16 ± 1.24 ($n = 391$)	0.13
T2 Hyperintensity ($n = 446$)	49.2% ($n = 32/65$)	75.6% ($n = 288/381$)	**<0.0001**
T1 Hypointensity ($n = 422$)	9.7% ($n = 6/62$)	21.1% ($n = 76/360$)	**0.036**
Levels of T2 Hyperintensity	0.82 ± 0.98 ($n = 65$)	1.3 ± 1.11 ($n = 381$)	**0.001**

NDI, Neck Disability Index; MRI, magnetic resonance imaging; mJOA, modified Japanese Orthopaedic Association scale; PCS, physical component summary; MCS, mental component summary.

On MRI, patients with GICs less commonly exhibited signal intensity changes (T2 hyperintensity, 49.2% vs. 75.6%, $p < 0.001$; T1 hypointensity, 9.7% vs. 21.1%, $p = 0.036$) and had a lower number of T2 hyperintensity levels (0.82 ± 0.98 vs. 1.3 ± 1.11, $p = 0.001$) than patients without GICs. However, there were no differences in the number of compressed levels or the prevalence of combined anterior–posterior compression.

There were no differences in neurological or functional outcomes at 2-year follow-up between patients with or without GICs (Tables 2 and 3).

Table 2. Surgical outcome at 2-years follow-up.

Outcome at 2-years (Mean Difference)	Gastrointestinal Comorbidity	Non-Gastrointestinal Comorbidity	*p*-Value
mJOA	2.96 ± 2.72 (*n* = 112)	2.66 ± 2.93 (*n* = 575)	0.30
Nurick	−1.55 ± 1.39 (*n* = 113)	−1.39 ± 1.47 (*n* = 586)	0.30
NDI	−10.84 ± 20.78 (*n* = 93)	−12.03 ± 19.04 (*n* = 512)	0.59
SF-36 PCS	5.95 ± 10.84 (*n* = 108)	5.64 ± 10.42 (*n* = 568)	0.78
SF-36 MCS	4.61 ± 15.66 (*n* = 108)	5.69 ± 13.74 (*n* = 568)	0.50

Table 3. Outcomes comparing GIC and non-GIC subgroups using linear mixed effects modeling.

Time	Change from Baseline *		Difference in Change, GIC vs. No GIC [†]	*p*-Value
	No GIC	GIC		
mJOA				
Baseline	12.48	12.74		
1 year.	2.56	2.65	0.08 (-0.51 to 0.68)	0.787
2 year.	2.67	3.07	0.40 (-0.21 to 1.01)	0.201
Nurick				
Baseline	3.28	3.05		
1 year.	−1.40	−1.47	−0.07 (−0.38 to 0.24)	0.661
2 year.	−1.42	−1.65	−0.23 (−0.55 to 0.09)	0.164
NDI				
Baseline	38.23	46.31		
1 year.	−11.76	−13.36	−1.60 (−6.11 to 2.91)	0.488
2 year.	−12.36	−10.65	1.71 (−2.90 to 6.33)	0.468
SF-36 PCS				
Baseline	34.65	32.80		
1 year.	6.16	6.83	0.67 (−1.58 to 2.92)	0.561
2 year.	5.59	6.51	0.93 (−1.38 to 3.23)	0.432
SF-36 MCS				
Baseline	40.57	38.47		
1 year.	6.26	5.89	−0.37 (−3.35 to 2.61)	0.807
2 year.	5.94	4.52	−1.42 (−4.48 to 1.63)	0.361

* Values are reported as least-squares means of change in outcome scores from baseline at each follow-up time point. Baseline scores are reported as mean values for each treatment group. [†] Values are reported as difference in means (95% CI). Confidence intervals have not been adjusted for multiplicity.

4. Discussion

Gastrointestinal conditions are common in the elderly, and therefore, it is not surprising that GICs were prevalent in 16% of patients with DCM. What is clear from this study is that patients with GICs represent a unique cohort that is quite different from the typical DCM patient: (1) they are more commonly female, despite the fact that prevalence of DCM among males is greater among reported studies (1), (2) almost a third of patients have psychiatric comorbidities, a much higher prevalence than otherwise expected, (3) patients had a large discrepancy between their general health measure score and NDI vs. neurological measures, showing significantly increased general health and neck disability but milder neurological impairment, and (4) GIC patients showed significantly lower MRI evidence of cord injury, despite having only subtle differences in neurological function. Given the large sample of patients with GICs ($n = 121$) in the cohort and the substantial deviance of clinical presentation in multiple dimensions (i.e., demographic, general health, neck pain, and objective findings of neurologic injury), it is clear that the presence of GICs is influential, but due to its broad categorization, it is challenging to account for specific factors.

Generally, GICs can result in a number of potential conditions, including anemia (due to blood loss), as well as malnutrition and vitamin deficiencies (due to GI resections or inflammatory conditions) that are essential to spinal cord function [6,7,22]. Anemia is usually easily identified through preoperative screening, and its preoperative presence should be managed prior to surgery to avoid complications. Indeed, it has been previously shown using the NSQIP database that preoperative anemia is an independent risk factor for complications, the need for perioperative blood transfusion, return to the operating room, and extended length of stay after cervical surgery [12,23]. From a different perspective, it has also been proposed through animal studies that spinal cord compression may result in irregular nervous stimulation of the stomach, a phenomenon termed neck-stomach syndrome [24]. However, this connection remains largely unexplored.

The potential role of nutritional or vitamin deficiencies in DCM has not been adequately investigated, and therefore, it is unclear how these patients would present. In general, it is known that lack of nutritional factors is contributory to the health of intervertebral discs, as the avascular nature of discs and reliance on diffusion renders them susceptible to injury due to undernutrition. In particular, nutritional levels must exceed a critical threshold for the cells to remain viable and active [25].

Two nutritional factors that may be specifically relevant to neurological function include B12 deficiency and copper deficiency (hypocupremia). Deficiency of either of these can result in both myelopathy and anemia [6,7]. Copper deficiency is rare, typically manifests due to high zinc intake, gastric resection, and malabsorption, and in a majority of cases treatment does not reverse myelopathic injury [6]. In contrast, B12 deficiency is much more common: It has been estimated that subclinical or clinical deficiency exists in up to 20% of elderly patients [26]. Further, clinical manifestation of B12 deficiency can mimic DCM and present with T2 cord hyperintensity. B12 deficiency has been reported to occur concomitantly with DCM [27–29], as well as in patients with suspected DCM—but underlying SACD—who experienced a resolution of symptoms after B12 administration [9,10,30]. B12 deficiency is most commonly due to pernicious anemia, bowel resection, inflammatory bowel disorders, liver disease, or gastric atrophy [31,32]. Unfortunately, without lab work to corroborate this, this remains speculative. However, B12 deficiency is also known to cause cognitive impairment and neuropsychiatric disease [33] and could be responsible for the high level of psychiatric comorbidities observed amongst patients with GICs in the present analysis. The relationship between psychiatric and gastrointestinal comorbidities, as well as other somatic symptoms such as back pain, has been previously reported [34,35]. For example, a recent study on irritable bowel syndrome concluded that psychiatric factors could contribute to predisposition, precipitation, and perpetuation of IBS symptoms [36]. Such findings suggest a potential explanation for the significantly different levels of neck disability between the two groups, as it is plausible that a higher rate of psychiatric comorbidities contributed to the higher rate of non-objectifiable symptoms. It also suggests that perhaps the high level of psychiatric symptoms is the reason for this different population clinical phenotype.

Overall, the findings suggest that patients with GICs were less commonly severely neurologically impaired. This is evidenced by the lower prevalence of objective upper motor signs (Babinski's reflex, hyperreflexia) and MRI evidence typical of more severely impaired patients (T1 hypointensity). Despite these and a marginally lower Nurick grade, there was no statistically significant difference in surgical outcomes between patients with or without GICs.

Limitations

A clear limitation to this study is the nonspecific nature of having classified patients into a single group of gastrointestinal comorbidities. It would have been preferable to know specific diagnoses;

however, these data were not available. Furthermore, given that the main study was not focused on gastrointestinal disease, we may have not captured an accurate population prevalence. Due to this, caution needs to be taken in interpreting the results, as false positive relationships are possible. Further, we have hypothesized that the unique differences observed here are possibly due to nutritional deficiencies; however, further work is needed to corroborate this. Lastly, because MRI data were derived from multiple global sites, there was no standardized protocol used to obtain MRIs.

5. Conclusions

Patients with GICs represent a unique cohort that is different from typical DCM patients: (1) they are more commonly female, (2) almost a third of patients have psychiatric comorbidities, and (3) they have worse general health and NDI findings, but less severe neurological deficits and MRI evidence of neurological impairment. This constellation of symptoms is considerably different than those typically observed in DCM; it is therefore plausible that nutritional factors that frequently manifest in elderly patients may contribute to this unique observation.

Author Contributions: Conceptualization, A.N.; Methodology, A.N. and J.H.B.; Validation, A.N. and J.H.B.; Formal Analysis, A.N., J.H.B., S.K. and H.R.-K.; Investigation, A.N. and M.G.F.; Resources, A.N., J.S.C., E.T., K.S. and M.G.F.; Data Curation, A.N. and J.H.B.; Writing-Original Draft Preparation, A.N., J.H.B. and K.P.; Writing-Review & Editing, All authors; Supervision, J.S.C., E.T., K.S. and M.G.F.; Project Administration, A.N., J.H.B. and K.P.; Funding Acquisition, A.N. and M.G.F. All authors have read and agreed to the published version of the manuscript.

Acknowledgments: The authors would like to acknowledge AOSpine for funding the initial AOSpine CSM-NA and CSM-I studies. MGF would like to acknowledge support from the Halbert Chair in Neural Repair and Regeneration and the DeZwirek Family Foundation.

Conflicts of Interest: The authors declare no conflict of interest.

References

1. Nouri, A.; Tetreault, L.; Singh, A.; Karadimas, S.; Fehlings, M. Degenerative Cervical Myelopathy: Epidemiology, Genetics and Pathogenesis. *Spine* **2015**, *40*, E675–E693. [CrossRef]
2. Nouri, A.; Martin, A.R.; Kato, S.; Reihani-Kermani, H.; Riehm, L.E.; Fehlings, M.G. The Relationship between MRI Signal Intensity Changes, Clinical Presentation, and Surgical Outcome in Degenerative Cervical Myelopathy: Analysis of a Global Cohort. *Spine J.* **2017**, *17*, S133–S134. [CrossRef]
3. Nouri, A.; Tetreault, L.; Dalzell, K.; Zamorano, J.J.; Fehlings, M.G. The Relationship between Preoperative Clinical Presentation and Quantitative Magnetic Resonance Imaging Features in Patients with Degenerative Cervical Myelopathy. *Neurosurgery* **2016**, *80*, 121–128. [CrossRef] [PubMed]
4. Harrop, J.S.; Naroji, S.; Maltenfort, M.; Anderson, D.G.; Albert, T.; Ratliff, J.K.; Ponnappan, R.K.; Rihn, J.A.; Smith, H.E.; Hilibrand, A.; et al. Cervical myelopathy: A clinical and radiographic evaluation and correlation to cervical spondylotic myelopathy. *Spine* **2010**, *35*, 620–624. [CrossRef] [PubMed]
5. Rhee, J.M.; Heflin, J.A.; Hamasaki, T.; Freedman, B. Prevalence of physical signs in cervical myelopathy: A prospective, controlled study. *Spine* **2009**, *34*, 890–895. [CrossRef] [PubMed]
6. Gabreyes, A.A.; Abbasi, H.N.; Forbes, K.P.; McQuaker, G.; Duncan, A.; Morrison, I. Hypocupremia associated cytopenia and myelopathy: A national retrospective review. *Eur. J. Haematol.* **2013**, *90*, 1–9. [CrossRef] [PubMed]
7. Stabler, S.P. Vitamin B12 deficiency. *N. Engl. J. Med.* **2013**, *368*, 2041–2042. [CrossRef] [PubMed]
8. Nouri, A.; Patel, K.; Montejo, J.; Nasser, R.; Gimbel, D.A.; Sciubba, D.M.; Cheng, J.S. The Role of Vitamin B12 in the Management and Optimization of Treatment in Patients with Degenerative Cervical Myelopathy. *Glob. Spine J.* **2018**, *9*, 331–337. [CrossRef]
9. Yokoyama, K.; Kawanishi, M.; Sugie, A.; Yamada, M.; Tanaka, H.; Ito, Y.; Kuroiwa, T. A Case of Subacute Combined Degeneration Caused by Vitamin B_2 Deficiency in a Cervical Spondylosis Surgery Referral. *No Shinkei Geka Neurol. Surg.* **2016**, *44*, 1059–1063.
10. Alonso, F.; Rustagi, T.; Schmidt, C.; Iwanaga, J.; Tubbs, R.S.; Chapman, J.R.; Fisahn, C. Subacute combined degeneration disguised as compressive myelopathy. *Spine Sch.* **2017**, *1*, 49–53. [CrossRef]

11. Nouri, A.; Matur, A.; Pennington, Z.; Elson, N.; Ahmed, A.K.; Huq, S.; Patel, K.; Jeong, W.; Nasser, R.; Tessitore, E.; et al. Prevalence of anemia and its relationship with neurological status in patients undergoing surgery for degenerative cervical myelopathy and radiculopathy: A retrospective study of 2 spine centers. *J. Clin. Neurosci.* **2020**, *72*, 252–257. [CrossRef] [PubMed]

12. Phan, K.; Wang, N.; Kim, J.S.; Kothari, P.; Lee, N.J.; Xu, J.; Cho SKPhan, K.; Wang, N.; Kim, J.S.; Kothari, P.; et al. Effect of Preoperative Anemia on the Outcomes of Anterior Cervical Discectomy and Fusion. *Glob. Spine J.* **2017**, *7*, 441–447. [CrossRef] [PubMed]

13. Lee, M.J.; Konodi, M.A.; Cizik, A.M.; Bransford, R.J.; Bellabarba, C.; Chapman, J.R. Risk factors for medical complication after spine surgery: A multivariate analysis of 1591 patients. *Spine J.* **2012**, *12*, 197–206. [CrossRef] [PubMed]

14. Puvanesarajah, V.; Jain, A.; Kebaish, K.; Shaffrey, C.I.; Sciubba, D.M.; De la Garza-Ramos, R.; Khanna, A.J.; Hassanzadeh, H. Poor nutrition status and lumbar spine fusion surgery in the elderly: Readmissions, complications, and mortality. *Spine* **2017**, *42*, 979–983. [CrossRef]

15. Fehlings, M.G.; Wilson, J.R.; Kopjar, B.; Yoon, S.T.; Arnold, P.M.; Massicotte, E.M.; Vaccaro, A.R.; Brodke, D.S.; Shaffrey, C.I.; Smith, J.S.; et al. Efficacy and safety of surgical decompression in patients with cervical spondylotic myelopathy: Results of the AOSpine North America prospective multi-center study. *J. Bone Jt. Surg.* **2013**, *95*, 1651–1658. [CrossRef]

16. Fehlings, M.G.; Ibrahim, A.; Tetreault, L.; Albanese, V.; Alvarado, M.; Arnold, P.; Barbagallo, G.; Bartels, R.; Bolger, C.; Defino, H.; et al. A global perspective on the outcomes of surgical decompression in patients with cervical spondylotic myelopathy: Results from the prospective multicenter AOSpine international study on 479 patients. *Spine* **2015**, *40*, 1322–1328. [CrossRef]

17. Ware, J.E., Jr.; Sherbourne, C.D. The MOS 36-item short-form health survey (SF-36). I. Conceptual framework and item selection. *Med. Care* **1992**, *30*, 473–483. [CrossRef]

18. Vernon, H.; Mior, S. The Neck Disability Index: A study of reliability and validity. *J. Manip. Physiol. Ther.* **1991**, *14*, 409–415.

19. Benzel, E.C.; Lancon, J.; Kesterson, L.; Hadden, T. Cervical laminectomy and dentate ligament section for cervical spondylotic myelopathy. *J. Spinal Disord.* **1991**, *4*, 286–295. [CrossRef]

20. Nurick, S. The pathogenesis of the spinal cord disorder associated with cervical spondylosis. *Brain* **1972**, *95*, 87–100. [CrossRef]

21. Nouri, A.; Martin, A.R.; Tetreault, L.; Nater, A.; Kato, S.; Nakashima, H.; Nagoshi, N.; Reihani-Kermani, H.; Fehlings, M.G. MRI analysis of the combined prospectively collected AOSpine North America and International Data: The Prevalence and Spectrum of Pathologies in a Global Cohort of Patients with Degenerative Cervical Myelopathy. *Spine* **2016**, *42*, 1058–1067. [CrossRef] [PubMed]

22. Guan, J.; Holland, C.M.; Ravindra, V.M.; Bisson, E.F. Perioperative malnutrition and its relationship to length of stay and complications in patients undergoing surgery for cervical myelopathy. *Surg. Neurol. Int.* **2017**, *8*, 307. [PubMed]

23. Phan, K.; Dunn, A.E.; Kim, J.S.; Capua, J.D.; Somani, S.; Kothari, P.; Lee, N.J.; Xu, J.; Dowdell, J.E.; Cho, S.K. Impact of Preoperative Anemia on Outcomes in Adults Undergoing Elective Posterior Cervical Fusion. *Global Spine J.* **2017**, *7*, 787–793. [CrossRef]

24. Song, X.-H.; Xu, X.-X.; Ding, L.-W.; Cao, L.; Sadel, A.; Wen, H. A preliminary study of neck-stomach syndrome. *WJG* **2007**, *13*, 2575. [CrossRef] [PubMed]

25. Huang, Y.-C.; Urban, J.P.; Luk, K.D. Intervertebral disc regeneration: Do nutrients lead the way? *Nat. Rev. Rheumatol.* **2014**, *10*, 561. [CrossRef] [PubMed]

26. Andrès, E.; Loukili, N.H.; Noel, E.; Kaltenbach, G.; Abdelgheni, M.B.; Perrin, A.E.; Noblet-Dick, M.; Maloisel, F.; Schlienger, J.L.; Blicklé, J.F. Vitamin B12 (cobalamin) deficiency in elderly patients. *CMAJ* **2004**, *171*, 251–259. [CrossRef] [PubMed]

27. Xu, Y.; Chen, W.; Jiang, J. Cervical spondylotic myelopathy with vitamin B deficiency: Two case reports. *Exp. Ther. Med.* **2013**, *6*, 943–946. [CrossRef]

28. Miyazaki, T.; Sudo, H.; Hiratsuka, S.; Iwasaki, N. Cervical spondylotic myelopathy with subacute combined degeneration. *Spine J.* **2014**, *14*, 381–382. [CrossRef]

29. Haghighi, S.S.; Zhang, R.; Stein, D. Cervical myelopathy due to chronic vitamin B12 deficiency or herniated cervical disc or both. *Electromyogr. Clin. Neurophysiol.* **2003**, *43*, 443–447.

30. Li, J.; Zhang, L.; Zhang, Y.; Deng, B.; Bi, X. Misdiagnosis of spinal subacute combined degeneration in a patient with elevated serum B12 concentration and sensory deficit level. *Neurol. Sci.* **2016**, *37*, 1577–1578. [CrossRef]
31. Nielsen, M.J.; Rasmussen, M.R.; Andersen, C.B.; Nexo, E.; Moestrup, S.K. Vitamin B12 transport from food to the body's cells—A sophisticated, multistep pathway. *Nat. Rev. Gastroenterol. Hepatol.* **2012**, *9*, 345–354. [CrossRef] [PubMed]
32. Hannibal, L.; Lysne, V.; Bjørke-Monsen, A.L.; Behringer, S.; Grünert, S.C.; Spiekerkoetter, U.; Jacobsen, D.W.; Blom, H.J. Biomarkers and Algorithms for the Diagnosis of Vitamin B12 Deficiency. *Front. Mol. Biosci.* **2016**, *3*, 27. [CrossRef]
33. Reynolds, E. Vitamin B12, folic acid, and the nervous system. *Lancet Neurol.* **2006**, *5*, 949–960. [CrossRef]
34. Walker, E.A.; Katon, W.J.; Jemelka, R.P.; Roy-Bryne, P.P. Comorbidity of gastrointestinal complaints, depression, and anxiety in the Epidemiologic Catchment Area (ECA) Study. *Am. J. Med.* **1992**, *92*, 26S–30S. [CrossRef]
35. Haug, T.T.; Mykletun, A.; Dahl, A.A. The association between anxiety, depression, and somatic symptoms in a large population: The HUNT-II study. *Psychosom Med.* **2004**, *66*, 845–851. [CrossRef] [PubMed]
36. Stasi, C.; Caserta, A.; Nisita, C.; Cortopassi, S.; Fani, B.; Salvadori, S.; Pancetti, A.; Bertani, L.; Gambaccini, D.; de Bortoli, N.; et al. The complex interplay between gastrointestinal and psychiatric symptoms in irritable bowel syndrome: A longitudinal assessment. *J. Gastroenterol. Hepatol.* **2019**, *34*, 713–719. [CrossRef]

Journal of
Clinical Medicine

Article

Validating the Transformation of PROMIS-GH to EQ-5D in Adult Spine Patients

Shreyas Panchagnula [1,†], Xin Sun [1,†], Julio D. Montejo [1,2], Aria Nouri [1,3,4], Luis Kolb [1], Justin Virojanapa [1,5], Joaquin Q. Camara-Quintana [1], Samuel Sommaruga [1,4], Kishan Patel [1], Nikita Lakomkin [6], Khalid Abbed [1] and Joseph S. Cheng [1,3,*]

1   Department of Neurosurgery, Yale School of Medicine, New Haven, CT 06511, USA
2   Dartmouth Hitchcock Medical Center, Lebanon, NH 03756, USA
3   Department of Neurosurgery, University of Cincinnati, Cincinnati, OH 45267, USA
4   Department of Neurosurgery, Geneva University Hospitals, 1205 Geneva, Switzerland
5   Donald and Barbara Zucker School of Medicine at Hofstra/Northwell, Hempstead, NY 11549, USA
6   Icahn School of Medicine at Mount Sinai, New York, NY 10029, USA
*   Correspondence: chengj6@ucmail.uc.edu; Tel.: +1-513-558-35562
†   Authors contributed equally.

Received: 1 August 2019; Accepted: 12 September 2019; Published: 20 September 2019

Abstract: Spinal disorders and associated interventions are costly in the United States, putting them in the limelight of economic analyses. The Patient-Reported Outcomes Measurement Information System Global Health Survey (PROMIS-GHS) requires mapping to other surveys for economic investigation. Previous studies have proposed transformations of PROMIS-GHS to EuroQol 5-Dimension (EQ-5D) health index scores. These models require validation in adult spine patients. In our study, PROMIS-GHS and EQ-5D were randomly administered to 121 adult spine patients. The actual health index scores were calculated from the EQ-5D instrument and estimated scores were calculated from the PROMIS-GHS responses with six models. Goodness-of-fit for each model was determined using the coefficient of determination (R^2), mean squared error (MSE), and mean absolute error (MAE). Among the models, the model treating the eight PROMIS-GHS items as categorical variables (CAT_{Reg}) was the optimal model with the highest R^2 (0.59) and lowest MSE (0.02) and MAE (0.11) in our spine sample population. Subgroup analysis showed good predictions of the mean EQ-5D by gender, age groups, education levels, etc. The transformation from PROMIS-GHS to EQ-5D had a high accuracy of mean estimate on a group level, but not at the individual level.

Keywords: EQ-5D; PROMIS; spine; transformation; quality of life; patient outcomes; validation

---

1. Introduction

High costs associated with surgical treatment of spine disorders demand a larger role for cost-utility analyses of treatment options. Amidst socioeconomic limitations and finite resources, spinal disorders occur at a high frequency, incur high costs for the healthcare system, and are treated with a heterogeneity of interventions. According to the 2010 Global Burden of Disease Study, low back pain had the greatest number of years lost to disability out of 291 conditions studied [1,2] and the annual direct costs of care provided for patients with spine disorders has been estimated at $90 billion [3]. Low back pain in particular presents a unique challenge, as there are numerous treatment modalities available whose comparative efficacy and value have not been fully substantiated [2].

Measuring the value of an intervention necessitates the use of a health utility score that encapsulates the health status, or patient-perceived overall health, at any given moment. Health status measures (HSMs) generally fall into two categories: (1) profile-based measures, such as the Patient-Reported

Outcomes Measurement Information System (PROMIS) [4]; and (2) preference-based measures, such as EuroQol 5-Dimension (EQ-5D) [5]. Profile-based measures characterize health status by assigning a score to each of multiple domains of health. Preference-based measures characterize health status by providing a single utility score from multiple domains of health. The utility score, based on valuations of different health states, is central to estimation of quality-adjusted life years (QALY), cost-utility analysis, cost-effectiveness of interventions, and quantitation of health outcomes [2,6].

Many health status measures have been designed for generic or disease-specific use [7,8]. In 1990, EuroQol developed the EQ-5D three-level survey (EQ-5D-3L, abbreviated as EQ-5D below), a preference-based HSM with two parts: (1) a descriptive survey with five questions assessing five dimensions of health; and (2) a visual analog scale that permits a numeric self-assessment of general health [5]. Responses to the descriptive survey yields a health utility index score.

In 2007, the National Institute of Health (NIH) developed PROMIS Global Health Survey (PROMIS-GHS), a standardized, self-reported profile-based HSM with 10 self-reported global health items that summarize general perceptions of health [4]. This survey is freely available for public use and is increasingly adopted in clinical settings. However, economic analyses have been classically performed using other preference-based measures, including EQ-5D.

With an increased desire to determine the value of health care and increase in HSMs, there is a growing interest to correlate different HSMs. In 2009, Revicki et al. facilitated a conversion from PROMIS-GHS to EQ-5D index scores using generic United States (US) population data [9]. Since then, many clinical studies have used this model (REV_{Reg}) when evaluating health outcomes of surgical and medical interventions [10–12]. While effective, such a conversion faces challenges and requires validation for specific patient populations or diseases. Furthermore, design of the model itself and its parameters can be optimized.

For instance, in 2017, Thompson et al. proposed new models to optimize REV_{Reg} using linear and equipercentile equating [13]. Linear and equipercentile equating are linking techniques that, after predicting scores, assign profile-based responses to preference-based scores by aligning score distributions of the two scales. Using Revicki et al.'s original data set, they recreated Revicki et al.'s regression model (REV_{Reg}), applied linear equating to REV_{Reg} (REV_{LE}), and applied equipercentile equating to REV_{Reg} (REV_{equip}). In a similar fashion, they created three models by treating the score as categorical variables (CAT_{Reg}, CAT_{LE}, CAT_{equip}) for a total of six models. They performed external validation of these models on a neurologic disease cohort from Cleveland Clinic.

In this study, we compared these six models in a cohort of adult spine patients to assess their ability to map PROMIS-GHS to EQ-5D in the spinal population.

2. Experimental Section

2.1. Surveys

A short demographics form was used to obtain gender, age, race/ethnicity, education, medical history, and spine diagnosis of participants.

The PROMIS Global Health survey includes ten global health items to assess overall health: (1) general health, (2) quality of life, (3) physical health, (4) mental health, (5) social satisfaction, (6) physical activities, (7) pain, (8) fatigue, (9) social activities, and (10) emotional distress. Every item except the pain item is rated on a numeric five-level scale (1 representing poor and 5 representing excellent); the pain item is scored from 0 to 10, where 0 indicates no pain and 10 indicates the worst imaginable pain. The pain item is then recoded to a five-level scale, and the fatigue and emotional problem item is recoded such that a high score represents better health status. Individual global item scores from completed PROMIS surveys were used to calculate estimates of EQ-5D index scores.

The EQ-5D is a preference-based instrument designed to measure generic health status across five dimensions of health: (1) mobility, (2) self-care, (3) usual activities, (4) pain/discomfort, and (5) anxiety/depression, with three response levels (no problems, some problems, extreme problems) [14].

A unique EQ-5D health state is defined by combining one level from each of the five dimensions, and each health state corresponds to a health index ranging from −0.109 to 1.0, with greater scores correlating to better overall health [15]. This index was calculated for every completed EQ-5D survey according to the valuations developed by Shaw et al. and derived from a large scale survey of the US general population [15]. The single visual analogue scale component of EQ-5D (EQ-5D VAS) was obtained but not evaluated in this study. Permission to use EQ-5D was granted by the EuroQol Group.

2.2. Study Design and Participants

This study was primarily conducted in the adult spine clinics of the three neurosurgeons (K.A., J.S.C, and L.K) at Yale University School of Medicine in New Haven, CT, with Institutional Review Board approval. Figure 1 illustrates the design of the study. In these clinics, 146 adult (>18 years of age) spine patients were recruited in 2017 as they entered the clinic with voluntary consent regardless of their clinical status (pre-operative, post-operative, or non-operative). Three forms were administered in paper to these patients: a demographics short form, PROMIS-GHS, and EQ-5D. PROMIS-GHS and EQ-5D were administered in random order. Completion of these two survey components was essential for obtaining an EQ-5D index and corresponding index estimates from PROMIS Global Health items. Out of 146 patients, complete survey responses were obtained from 121 patients.

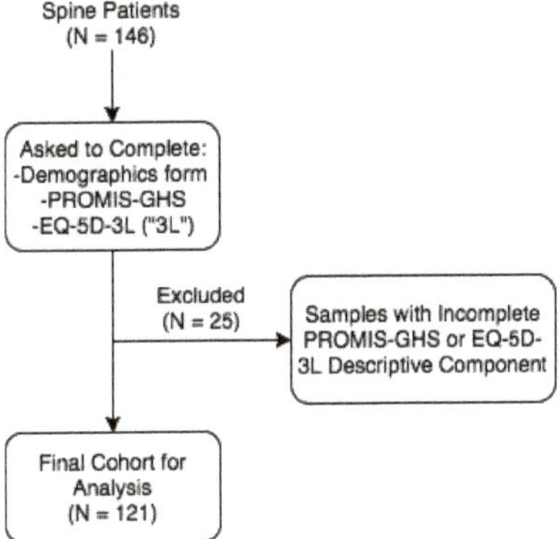

Figure 1. Sample Selection. PROMIS-GHS: Patient-Reported Outcomes Measurement Information System Global Health Survey; 5Q-5D: EuroQol 5-Dimension; 5Q-5D-3L; EQ-5D three-level survey.

2.3. Models Tested in the Study

REV$_{Reg}$: This model was developed in 2009 by applying ordinary least squares (OLS) regression on the PROMIS Wave 1 Sample (i.e., the sample used by Revicki et al.) [13,16] to predict EQ-5D index scores from PROMIS-GHS items. This model uses eight out of 10 PROMIS-GHS items in its algorithm (excluding responses to general health and social satisfaction) and treats these items as continuous variables.

REV$_{LE}$: This model is the result of applying linear equating, a method of linking, to REV$_{Reg}$. While regression models aim to predict preference-based scores from profile-based responses, linking models align score distributions of observed and predicted scores to establish a scale that provides an equivalent preference-based score for each set of profile-based responses. Linear equating is applied to REV$_{Reg}$ with the following equation:

$$Y_{LE} = \mu_Y + \frac{\sigma_Y}{\sigma_{Y_R}}\left(Y_R - \mu_{Y_R}\right) \tag{1}$$

where Y_{LE} is the estimated value from linear equating, μ_Y and σ_Y are the mean and standard deviation of the observed EQ-5D scores from the PROMIS Wave 1 Sample, respectively, and μ_{Y_R} and σ_{Y_R} are the mean and standard deviation of the predicted EQ-5D scores from REV$_{Reg}$, respectively.

REV$_{equip}$: This model was developed by applying equipercentile equating to REV$_{Reg}$. Equipercentile equating is a linking method that matches the cumulative distribution functions of observed scores and predicted scores from REV$_{Reg}$ using smoothing functions or nonparametric techniques.

CAT$_{Reg}$: This model was implemented in 2017 by Thompsons et al. Like REV$_{Reg}$, this model utilizes OLS regression on the PROMIS Wave 1 sample to predict EQ-5D index scores from eight PROMIS-GHS items. Unlike REV$_{Reg}$, CAT$_{Reg}$ treats these items as categorical variables.

CAT$_{LE}$: This model is the result of applying linear equating to CAT$_{Reg}$.

CAT$_{equip}$: This model was developed by applying equipercentile equating to CAT$_{Reg}$.

2.4. Statistical Analysis

Statistical analyses were conducted in R Studio [17]. Responses to each of the 121 completed EQ-5D surveys were utilized to calculate an EQ-5D index score according to the valuations developed by Shaw et al. [15]. Estimates of the EQ-5D index scores from PROMIS Global Health Item responses were obtained by applying the six models developed by Revicki et al. and Thompson et al. (REV$_{Reg}$, REV$_{LE}$, REV$_{equip}$, CAT$_{Reg}$, CAT$_{LE}$, CAT$_{equip}$) [9,13].

The goodness of fit for each model in our sample of patients was measured with the Pearson correlation coefficient (r), coefficient of Determination (R^2), mean squared error (MSE), and mean absolute error (MAE). Correlation r measures the strength of the linear relationship. Higher absolute values indicate stronger linear correlations. R^2 demonstrates how much variance could be explained by the regression model. The mean squared error (MSE) and mean absolute error (MAE) were measured to examine the scale of difference between each estimate and observed value. Models with lower MSE or MAE have better predictions.

In addition, comparisons of actual EQ-5D scores and optimal estimates were performed by subgroups, such as gender, age groups, ethnicity, education, and spine diagnosis. According to Luo et al., 0.04 was recommended as the minimal clinically important difference of a EQ-5D utility score with a scale from -0.109 to 1 [18]. If the mean difference is less than 0.04, we consider it is an accurate estimate of the mean.

However, good linear correlation does not always imply good agreement. In order to evaluate the transformation on an individual level, the Bland–Altman assessment of agreement was conducted. It could visually show the difference between actual and estimated scores of each patient. Histograms of the observed EQ-5D scores and estimates from each model were also plotted to show distributions of scores.

3. Results

3.1. Demographic Characteristics

Table 1 contains the demographics of the experimental cohort of adult spine patients. Our cohort of 121 patients had an average age of 59 years, was 59% female, and had a majority with Caucasian race/ethnicity. Highest level of education in these patients ranged from less than high school (4%) to advanced college degree (17%), with 33% completing high school, 31% having some college or associate's degree, and 14% having a bachelor's degree. Patients had a variety of conditions in their medical histories, including cancer, lung disease, psychiatric illness, heart disease, rheumatologic disease, central nervous system (CNS) disorders, and liver/kidney disease.

Table 1. Demographic and clinical characteristics of survey participants.

Characteristic	Spine Patients (*N* = 121)
Age, mean ± SD	59 ± 13
Gender, *n* (%)	
Female	71 (59)
Male	49 (40)
Race/Ethnicity, *n* (%)	
Caucasian American	94 (77)
African American	13 (11)
Hispanic American	9 (7)
Caucasian American and Hispanic American	1 (1)
Asian American	1 (1)
Caucasian American and Native American	1 (1)
Highest Level of Education, *n* (%)	
Advanced Degree	21 (17)
Bachelor's Degree	17 (14)
Some College or Associate's Degree	38 (31)
High School Completion	40 (33)
Less than High School	5 (4)
Medical History, *n* (%)	
Psychiatric Illness	33 (27)
Lung Disease	30 (25)
Heart Disease	27 (22)
Cancer/Tumor	25 (21)
CNS disorders	18 (15)
Rheumatologic Disease	17 (14)
Liver/Kidney Disease	11 (9)
Spine Diagnosis, *n* (%)	
Stenosis	35 (29)
Radiculopathy	14 (12)
Myelopathy	13 (11)
Deformity	12 (10)
Disc Herniation	5 (4)
Spondylolisthesis	5 (4)
Fracture	3 (2)
Tumor	3 (2)
Pseudoarthrosis	1 (1)

The cohort of this study had demographics comparable to the sample of the generic US population studied by Revicki et al. [9] and the neurologic disease cohort studied by Thompson et al. [13]. Unlike Revicki et al. and Thompson et al., however, all sample subjects had spine diagnoses, including cervical and lumbar stenosis (most common), deformity, myelopathy, radiculopathy, spondylolisthesis, fracture, tumor, and pseudoarthrosis. The specificity of spine diagnosis distinguishes the cohort of this study from the general cohort of Revicki's study.

3.2. Statistical Analysis

Table 2 presents the metrics used to assess the models applied to our sample. The estimated score in the CAT_{Reg} model (0.60) was closest to the observed EQ-5D index scores (0.62). The mean difference was 0.012 (95% CI, −0.012–0.036, $p = 0.3144$), which indicated no significant difference between actual EQ-5D score and CAT_{Reg} estimates. All other estimates were significantly different using the paired t-test. The R^2 values for all six models ranged between 0.54 and 0.59. Pearson correlation coefficients were all above 0.7, showing strong linear correlation. Of the six models, CAT_{Reg} had the highest R^2 (0.59) and lowest MSE (0.02) and MAE (0.11). Thus, CAT_{Reg} is the optimal model among them.

Table 2. Mean (standard deviation (SD)) of actual and estimated EQ-5D Index Scores, R^2 values, correlation coefficients, mean squared errors (MSE), and mean absolute errors (MAE) for models in the spine patient sample (N = 121).

	Mean (SD)	R^2	r	MSE	MAE
Actual	0.62 (0.21)				
REV_{Reg}	0.57 (0.10)	0.57	0.76	0.02	0.13
REV_{LE}	0.56 (0.17)	0.57	0.76	0.02	0.12
REV_{equip}	0.54 (0.22)	0.57	0.76	0.03	0.12
CAT_{Reg}	0.60 (0.18)	0.59	0.77	0.02	0.11
CAT_{LE}	0.56 (0.22)	0.59	0.77	0.02	0.12
CAT_{equip}	0.56 (0.23)	0.54	0.73	0.03	0.13

In order to investigate the accuracy of CAT_{reg} model predictions, subgroup analysis was also performed, shown in Table 3. Within most subgroups, the mean difference was less than 0.04 (the minimal clinically important difference of EQ-5D score), which means the EQ-5D score could be accurately predicted using PROMIS-GHS. For example, the female spine patients' observed EQ-5D score was 0.62 and the estimate of CATreg was 0.60 (95% CI, 0.56–0.64), while the males' was 0.60 vs. 0.60 (95% CI, 0.55–0.66). Caucasian Americans had a higher average EQ-5D score (actual 0.64 vs. estimates 0.64) than other ethnicities (actual 0.52 vs. estimates 0.50). The actual score for different education level ranged from 0.53 to 0.70. Generally, the larger the group size, the better prediction was achieved. All the subgroups with more than 17 patients had a mean difference less than 0.04, which indicates this score transformation should be more appropriately used on a group level, instead of individual level.

Table 3. Comparison of actual EQ-5D scores and estimates of the CAT_{reg} model by subgroups.

	N	Actual EQ-5D Mean (SD)	CATreg Estimates Mean (SD)	Mean Difference
Gender				
Female	71	0.62 (0.20)	0.60 (0.16)	0.02
Male	49	0.60 (0.22)	0.60 (0.20)	0.00
Age groups, years				
18–45	17	0.59 (0.22)	0.54 (0.21)	0.05
46–65	63	0.60 (0.23)	0.59 (0.18)	0.01
65+	40	0.65 (0.16)	0.65 (0.15)	0.00
Ethnicity				
Caucasian American	94	0.64 (0.20)	0.64 (0.16)	0.00
Others	27	0.52 (0.23)	0.50 (0.18)	0.02
Highest education level				
Advanced degree	21	0.65 (0.16)	0.69 (0.18)	−0.04
Bachelor's degree	17	0.70 (0.16)	0.62 (0.16)	0.08
Some college or associate's degree	38	0.61 (0.22)	0.61 (0.15)	0.00
High school completion	40	0.58 (0.23)	0.56 (0.19)	0.02
Less than high school	5	0.53 (0.27)	0.48 (0.19)	0.05
Spine Diagnosis				
Stenosis	35	0.65 (0.20)	0.61 (0.17)	0.04
Other	22	0.59 (0.22)	0.58 (0.20)	0.01
Radiculopathy	14	0.65 (0.20)	0.63 (0.20)	0.03
Myelopathy	13	0.60 (0.24)	0.60 (0.21)	0.00
Deformity	12	0.62 (0.19)	0.61 (0.16)	0.01
Disc herniation	5	0.58 (0.27)	0.64 (0.25)	−0.06
Spondylolisthesis	5	0.50 (0.29)	0.48 (0.20)	0.02
Unknown	4	0.60 (0.18)	0.63 (0.14)	−0.03
Fracture	3	0.63 (0.06)	0.71 (0.02)	−0.08
Tumor	3	0.63 (0.06)	0.56 (0.07)	0.07
Herniated disc	2	0.31 (0.00)	0.50 (0.05)	−0.19
Pseudoarthrosis	1	0.44	0.72	−0.29

In order to investigate the prediction performance at an individual level, Bland-Altman analysis was conducted. Figure 2 demonstrated the mean residual was 0.01, with 95% limits of agreement between actual and CAT_{reg} estimated EQ-5D scores ranged from −0.25 to 0.27. It revealed that for a single patient, the variation from their actual score is huge and largely exceeded the minimal clinically important difference 0.04.

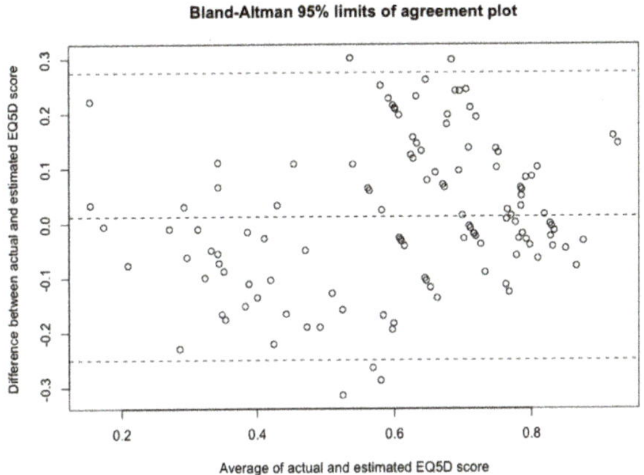

Figure 2. Bland-Altman agreement plot. X axis is the average score of the actual and estimated EQ-5D score of the CAT_{reg} model. Y axis is the difference between the two. Each dot represents a patient. The three dashed lines are upper 95% limits of agreements (mean + 1.96 SD), mean difference, and lower 95% limits of agreements (mean − 1.96 SD).

Figure 3 depicts histograms of the observed EQ-5D-3L scores and estimates from REVReg, REVLE, and REVequip in our sample. Figure 4 depicts histograms of the observed EQ-5D scores and estimates from CAT_{Reg}, CAT_{LE}, and CAT_{equip}. In both figures, the histograms of regression estimates and linear equating estimates resemble a normal distribution, while the histograms of the observed 3L scores and equipercentile equating estimates have a bimodal distribution. These histograms confirmed that the estimates from the transformation models are not a good match on an individual level.

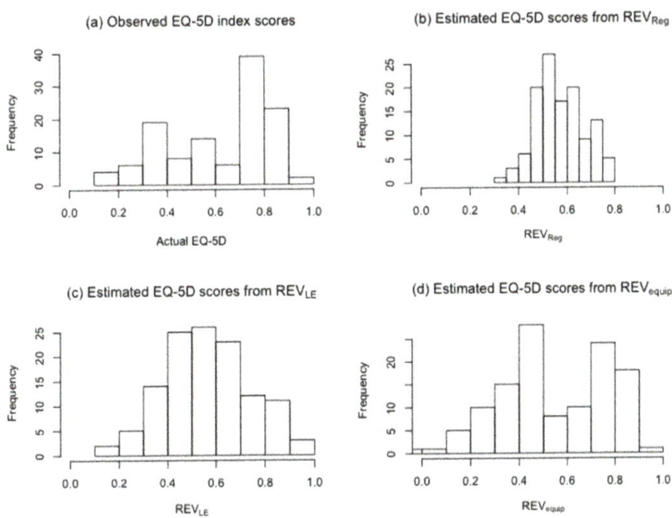

Figure 3. Histograms of observed EQ-5D index scores and estimates from REV_{Reg}, REV_{LE}, and REV_{equip}.

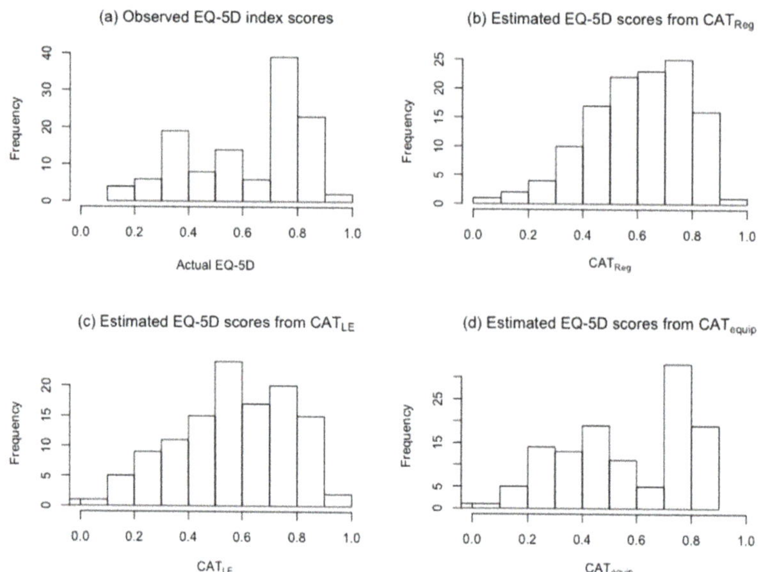

Figure 4. Histograms of observed EQ-5D index scores and estimates from CAT_{Reg}, CAT_{LE}, and CAT_{equip}.

4. Discussion

4.1. Validation and Technical Aspects

Our study assessed and compared six models that were developed in a generic sample to map PROMIS-GHS to EQ-5D in a specific sample of patients with spinal disorders. In our sample of patients with spinal disease, all six models achieved an R^2 greater than 0.5. According to Brazier et al., models that map to preference-based scores commonly achieve an R^2 of greater than 0.5 within the sample of model development [19]. R^2 as a measure of goodness-of-fit can determine how well the model explains the dataset it was estimated on. However, it did not show the scale of difference. In that regard, MSE and MAE can better assess mapping functions by indicating size of prediction errors [19]. So, we compared the models with consideration of all the goodness-of-fit indicators.

First, we agreed that treating PROMIS-GHS item scores 1 to 5 as categorical variables (CAT_{Reg}) performed better than treating them as continues variables (REV_{Reg}), with closer mean estimate (0.60 vs. 0.57, actual score = 0.62), higher R^2 (0.59 vs. 0.57), and lower MAE (0.11 vs. 0.13). However, unlike the recommendation of using equating technics used in the Thompson et al. article, in our spine sample population, the linear and equipercentile equating models (REV_{LE}, and REV_{equip}, CAT_{LE}, and CAT_{equip}) did not work well compared to the CAT_{Reg}. Thus, although all six models demonstrated adequate prediction ability, the CAT_{Reg} model is the optimal one for patients with spine disease.

Second, we recommend using this transformation from PROMIS-GHS to EQ-5D utility score on group-level mean estimates, not for individual prediction. From the subgroup analysis, it showed the accurate prediction (mean difference less than 0.04) was achieved in groups with more than 17 patients. To be more conservative, sample sizes of at least 30 patients are suggested for the good mean estimate of a EQ-5D score from PROMIS-GHS using the CAT_{Reg} model.

4.2. Utility of Health Care Measurement

HSMs have often been validated in patients with spinal disease before clinical application. For instance, Guilfoyle et al. validated the Medical Outcomes Study Short Form (SF-6, -12, -36), a general health outcome measure, in patients with lumbar disc prolapse, lumbar canal stenosis, and degenerative cervical myeloradiculopathy. This study found strong correlation between SF surveys and disease-specific measures such as the Roland Morris Disability Score (RMDS), Myelopathy Disability Index (MDI), and Hospital Anxiety and Depression Scales (HADS) [20,21]. Similarly, EQ-5D was assessed for its validity for use in spine surgery by comparison with the Oswestry Disability Index (ODI) in a study of patients who underwent lumbar spine surgery for degenerative disorders [21,22]. According to the study, EQ-5D and ODI were equal in assessment of health state, thus validating the use of EQ-5D in patients with spinal disorders.

The validation of EQ-5D and other HSM questionnaires in patients with spinal disorders paved the way for assessing the value of spinal interventions using health utility index scores. For instance, Witiw. et al. assessed the lifetime incremental cost-utility of surgical treatment for degenerative cervical myelopathy in a prospective observational cohort study by calculating health utility and QALYs from SF-6D [23]. Tosteson et al. used data from the Spine Patient Outcomes Research Trial (SPORT) to determine that lumbar discectomy was a clinically beneficial and cost-effective treatment of intervertebral disc herniation [24]. They also determined that spinal stenosis surgery was cost-effective but degenerative spondylolisthesis surgery was not cost-effective over a period of two years [25]. Conclusions from Tosteson et al. were based on the use of the EQ-5D index to obtain measures of QALY and incremental cost-effectiveness ratio.

Cost-utility studies of spinal interventions have also used estimation models to obtain health utility scores from other surveys. Qureshi et al. investigated the cost-effectiveness of anterior cervical discectomy and fusion (ACDF) and cervical disc replacement (CDR) as therapies for single-level cervical degenerative disc disease (DDD) [26]. To do this, the group used results of the 36-Item Short Form Health Survey (SF-36) from the ProDisc-C investigational device exemption study along with a model generated to estimate preference-based index scores from the Short Form-6 dimensions (SF-6D) (derived from a subsection of SF-36 items) [27].

Mapping PROMIS to EQ-5D can prove to be a powerful method of calculating health utility in economic cost-benefit studies. Along with its increased use by the NIH, PROMIS and its domain item banks allow flexibility in administration using either targeted short forms or computerized adaptive tests [4,9]. The importance of validating models such as those developed by Revicki et al. [9] and Thompson et al. [13] lies in assessing the clinical and economic utility of applying generic models to disease-specific populations, including those with spinal pathologies.

4.3. Clinical Implications

The findings in the paper indicate that PROMIS can act as a reasonable surrogate for EQ-5D. For hospitals or medical centers that have already collected PROMIS-GHS and do not have EQ-5D, they could use this transformation to estimate EQ-5D scores and then calculate the quality adjusted life-year for cost-effectiveness analysis. Based on previous reports and our data, it appears that CAT_{Reg} is the choice with the lowest error for patients with spinal disorders.

Measurement of health status not only assesses general cost-effectiveness of interventions but also provides the opportunity to assess individual patients longitudinally. Consequently, one can assess changes in conservative management, and treatment modalities can be altered in accordance to health status. Regular clinical usage of HSMs develops a general repository of health outcomes data that would otherwise come solely from research studies, potentially alleviating substantial costs for prospective research studies.

4.4. Limitations

One of the limitations of this study was the sample size of the cohort. Though the cohort in this study had a variety of spine pathologies, our sample size was limited to clinics in a single institution. The results may not represent the whole spine population. Second, we tried to create our own prediction model. However, only three out of 10 PROMIS-GHS items (general health, social satisfaction, pain) were significant predictors due to the limited sample size.

5. Conclusions

This study assesses and compares six models that map PROMIS-GHS to EQ-5D index values in a population of patients with spinal disorders. All six models demonstrate adequate and comparable predictive performance in our sample, thus validating their economic utility. Among the six models, the CATreg model is recommended for spine patients. That is, EQ-5D utility scores could be most accurately estimated by the linear combination of eight significantly correlated items from PROMIS-GHS, while scores 1 to 5 for each item is treated as a categorical variable. In addition, we suggest using this transformation model for group-based estimates, instead of for individual patient's EQ-5D score estimates. Validation studies of HSMs can lead to their application in cost-utility analyses.

Author Contributions: Conceptualization, S.P., X.S., J.D.M., A.N., L.K., J.V., J.Q.C.-Q., S.S., N.L. and J.S.C.; methodology, S.P., X.S., J.D.M., A.N., L.K., J.V., J.Q.C.-Q., N.L., and J.S.C.; software, S.P., X.S. and J.D.M.; formal analysis, S.P., X.S. and A.N.; investigation, S.P., X.S., J.D.M., A.N., L.K., J.V., S.S., K.A., and J.S.C.; data curation, S.P., X.S. and A.N.; writing—original draft preparation, S.P.; writing—review and editing, X.S., J.D.M., A.N., K.P. and J.S.C.; visualization, S.P., and X.S.; supervision, J.S.C. and A.N.; project administration, X.S., A.N., J.S.C.

Conflicts of Interest: The authors declare no conflict of interest.

References

1. Hoy, D.; March, L.; Brooks, P.; Blyth, F.; Woolf, A.; Bain, C.; Williams, G.; Smith, E.; Vos, T.; Barendregt, J.; et al. The global burden of low back pain: Estimates from the Global Burden of Disease 2010 study. *Ann. Rheum. Dis.* **2014**, *73*, 968–974. [CrossRef] [PubMed]

2. Resnick, D.K.; Tosteson, A.N.; Groman, R.F.; Ghogawala, Z. Setting the equation: Establishing value in spine care. *Spine* **2014**, *39*, S43–S50. [CrossRef] [PubMed]

3. Mroz, T.E.; McGirt, M.; Chapman, J.R.; Anderson, G.; Fehlings, M. More "why" and less "how": Is value-based spine care the next breakthrough? *Spine* **2014**, *39*, S7–S8. [CrossRef] [PubMed]

4. Cella, D.; Yount, S.; Rothrock, N.; Gershon, R.; Cook, K.; Reeve, B.; Ader, D.; Fries, J.F.; Bruce, B.; Rose, M. The Patient-Reported Outcomes Measurement Information System (PROMIS): Progress of an NIH roadmap cooperative group during its first two years. *Med. Care* **2007**, *45*, S3–S11. [CrossRef]

5. Brooks, R.; Rabin, R.; De Charro, F. (Eds.) *The Measurement and Valuation of Health Status Using EQ-5D: A European Perspective*; Springer Science & Business Media: Berlin, Germany, 2013.

6. Angevine, P.D.; Berven, S. Health economic studies: An introduction to cost-benefit, cost-effectiveness, and cost-utility analyses. *Spine* **2014**, *39*, S9–S15. [CrossRef]

7. Dakin, H. Review of studies mapping from quality of life or clinical measures to EQ-5D: An online database. *Health Qual. Life Outcomes* **2013**, *11*, 151. [CrossRef] [PubMed]

8. Sauerland, S.; Weiner, S.; Dolezalova, K.; Angrisani, L.; Noguera, C.M.; Garcia-Caballero, M.; Rupprecht, F.; Immenroth, M. Mapping utility scores from a disease-specific quality-of-life measure in bariatric surgery patients. *Value Health* **2009**, *12*, 364–370. [CrossRef]

9. Revicki, D.A.; Kawata, A.K.; Harnam, N.; Chen, W.H.; Hays, R.D.; Cella, D. Predicting EuroQol (EQ-5D) scores from the patient-reported outcomes measurement information system (PROMIS) global items and domain item banks in a United States sample. *Qual. Life Res.* **2009**, *18*, 783–791. [CrossRef]

10. Forsyth, A.L.; Witkop, M.; Lambing, A.; Garrido, C.; Dunn, S.; Cooper, D.L.; Nugent, D.J. Associations of quality of life, pain, and self-reported arthritis with age, employment, bleed rate, and utilization of hemophilia treatment center and health care provider services: Results in adults with hemophilia in the HERO study. *Patient Prefer. Adherence* **2015**, *9*, 1549–1560. [CrossRef]

11. Grosse, S.D.; Prosser, L.A.; Asakawa, K.; Feeny, D. QALY weights for neurosensory impairments in pediatric economic evaluations: Case studies and a critique. *Expert Rev. Pharmacoecon. Outcomes Res.* **2010**, *10*, 293–308. [CrossRef]

12. Hung, W.W.; Ross, J.S.; Farber, J.; Siu, A.L. Evaluation of the mobile acute care of the elderly (MACE) service. *JAMA Intern. Med.* **2013**, *173*, 990–996. [CrossRef] [PubMed]

13. Thompson, N.R.; Lapin, B.R.; Katzan, I.L. Mapping PROMIS Global Health Items to EuroQol (EQ-5D) Utility Scores Using Linear and Equipercentile Equating. *PharmacoEconomics* **2017**, *35*, 1167–1176. [CrossRef] [PubMed]

14. EuroQol Group. EuroQol—A new facility for the measurement of health-related quality of life. *Health Policy* **1990**, *16*, 199–208. [CrossRef]

15. Shaw, J.W.; Johnson, J.A.; Coons, S.J. US valuation of the EQ-5D health states: Development and testing of the D1 valuation model. *Med. Care* **2005**, *43*, 203–220. [CrossRef] [PubMed]

16. Cella, D. PROMIS 1 Wave 1. Harvard Dataverse: 2015. Available online: https://dataverse.harvard.edu/dataset.xhtml?persistentId=hdl:1902.1/21134 (accessed on 13 September 2019).

17. RStudio Team. *RStudio: Integrated Development for R*; RStudio, Inc.: Boston, MA, USA, 2016; Available online: http://www.rstudio.com/ (accessed on 13 September 2019).

18. Luo, N.; Johnson, J.A.; Coons, S.J. Using Instrument-Defined Health State Transitions to Estimate Minimally Important Differences for Four Preference-Based Health-Related Quality of Life Instruments. *Med. Care* **2010**, *48*, 365–371. [CrossRef]

19. Brazier, J.E.; Yang, Y.; Tsuchiya, A.; Rowen, D.L. A review of studies mapping (or cross walking) non-preference based measures of health to generic preference-based measures. *Eur. J. Health Econ.* **2010**, *11*, 215–225. [CrossRef]

20. Guilfoyle, M.R.; Seeley, H.; Laing, R.J. The Short Form 36 health survey in spine disease—Validation against condition-specific measures. *Br. J. Neurosurg.* **2009**, *23*, 401–405. [CrossRef]

21. McCormick, J.D.; Werner, B.C.; Shimer, A.L. Patient-reported outcome measures in spine surgery. *J. Am. Acad. Orthop. Surg.* **2013**, *21*, 99–107. [CrossRef]

22. Solberg, T.K.; Olsen, J.A.; Ingebrigtsen, T.; Hofoss, D.; Nygaard, O.P. Health-related quality of life assessment by the EuroQol-5D can provide cost-utility data in the field of low-back surgery. *Eur. Spine J.* **2005**, *14*, 1000–1007. [CrossRef]

23. Witiw, C.D.; Tetreault, L.A.; Smieliauskas, F.; Kopjar, B.; Massicotte, E.M.; Fehlings, M.G. Surgery for degenerative cervical myelopathy: A patient-centered quality of life and health economic evaluation. *Spine J.* **2017**, *17*, 15–25. [CrossRef]

24. Tosteson, A.N.; Skinner, J.S.; Tosteson, T.D.; Lurie, J.D.; Andersson, G.B.; Berven, S.; Grove, M.R.; Hanscom, B.; Blood, E.A.; Weinstein, J.N. The cost effectiveness of surgical versus nonoperative treatment for lumbar disc herniation over two years: Evidence from the Spine Patient Outcomes Research Trial (SPORT). *Spine* **2008**, *33*, 2108–2115. [CrossRef]

25. Tosteson, A.N.; Lurie, J.D.; Tosteson, T.D.; Skinner, J.S.; Herkowitz, H.; Albert, T.; Boden, S.D.; Bridwell, K.; Longley, M.; Andersson, G.B.; et al. Surgical treatment of spinal stenosis with and without degenerative spondylolisthesis: Cost-effectiveness after 2 years. *Ann. Intern. Med.* **2008**, *149*, 845–853. [CrossRef]

26. Qureshi, S.; Goz, V.; McAnany, S.; Cho, S.K.; Hecht, A.C.; Delamarter, R.B.; Fehlings, M.G. Health state utility of patients with single-level cervical degenerative disc disease: Comparison of anterior cervical discectomy and fusion with cervical disc arthroplasty. *J. Neurosurg. Spine* **2014**, *20*, 475–479. [CrossRef]

27. Brazier, J.; Roberts, J.; Deverill, M. The estimation of a preference-based measure of health from the SF-36. *J. Health Econ.* **2002**, *21*, 271–292. [CrossRef]

MDPI

St. Alban-Anlage 66

4052 Basel

Switzerland

Tel. +41 61 683 77 34

Fax +41 61 302 89 18

www.mdpi.com

Journal of Clinical Medicine Editorial Office

E-mail: jcm@mdpi.com

www.mdpi.com/journal/jcm